I0824827

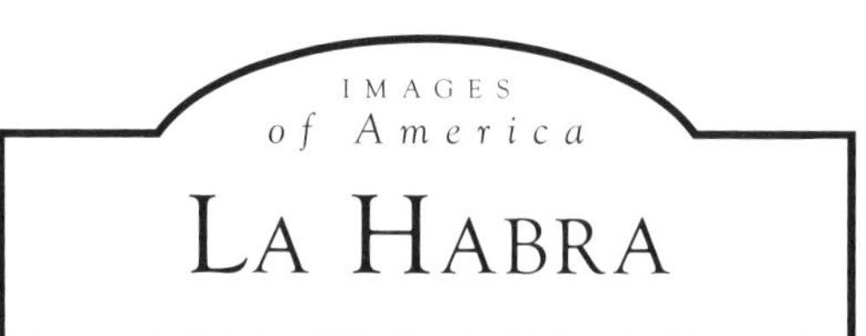
IMAGES
of America
LA HABRA

On the Cover: The second annual Old Settlers' Picnic was hosted by the Ladies' Mutual Improvement Club of La Habra at Martinez Grove in 1899. Participants included Willets J. Hole (standing at center rear), a land developer who is considered "the father of La Habra," and the Swiss settler families of Luehm (left front) and Leutwiler (right front). (Courtesy of the La Habra Historical Society.)

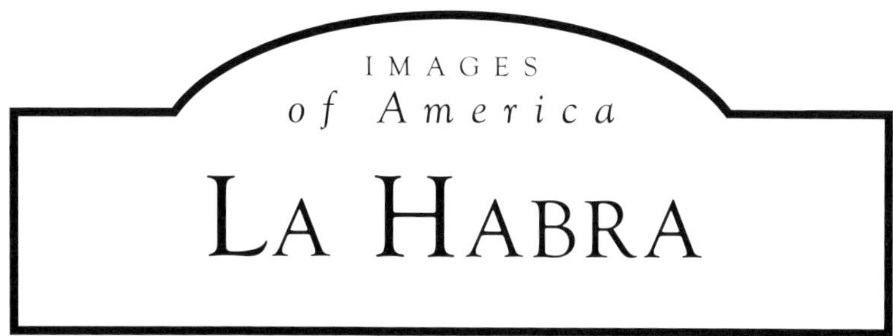

Lauren Blazey, Kimberly Albarian, and the
La Habra Centennial Celebration Committee

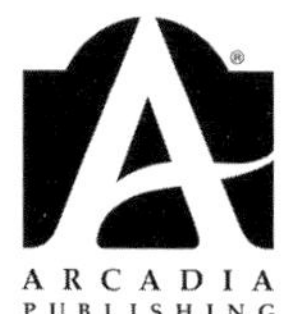

ISBN 9781-4671-6273-9

Published by Arcadia Publishing
Charleston, South Carolina

Printed in the United States of America

Library of Congress Control Number: 2025937195

For all general information, please contact Arcadia Publishing:
Telephone 843-853-2070
Fax 843-853-0044
E-mail sales@arcadiapublishing.com

Visit us on the Internet at www.arcadiapublishing.com

Contents

ACKNOWLEDGMENTS

We wish to extend a heartfelt thank you to the many individuals, families, and organizations who contributed their time, knowledge, and resources to this project. This book could not have been created without the generosity of those who shared photographs, documents, and personal stories to help preserve our history for future generations. A special thanks goes to the individuals who provided rare or fragile images and helped identify the people and places shown within these pages.

We are deeply grateful to the La Habra Centennial Celebration Committee, who helped lead and shape the city's 100th anniversary celebration. Their efforts ensured that our centennial year would be commemorated in a way that reflects both the city's proud history and the community spirit that continues to define La Habra. Serving on the committee were: Jess Badillo, Danitza Cardenas, Jim Coombs, Gina Cosylion, David DeLeon, Rose Espinoza, Kathy Felix, Gabby Garcia, Kandace Gutierrez, Aubrey Lebard, Annette Limon, Randy McMillan, Scott Miller, Mike Murphy, Debbie Musser, Vanessa Perkins, Norma Perez-Ochoa, Roy Ramsland, Denise Schmidt, Mary Schultz, Natalie Skullr-White, Dawn Stille, Carrie Surich, Jimmy Andreoli, and Karl Zener.

Special thanks also go to the La Habra Historical Museum and its president, Roy Ramsland, for their assistance in locating and preserving key photographs and to Kelly Fujio, director of community services, who had the vision to produce this book as a meaningful way to celebrate our rich history. We also thank the La Habra City Council, the City of La Habra, author and consultant Lauren Blazey, Autumn Dixon for her research contributions, and the countless community members who answered questions and helped verify historical details.

Finally, we want to thank everyone who spent many hours scanning, reviewing, and describing images. We are grateful to those who patiently loaned materials for digitizing. Your hard work has helped ensure that La Habra's centennial year will be remembered not only for its celebrations but also for preserving the stories and people who shaped the city's first 100 years.

Introduction

In the rancho days when vast herds of cattle and horses grazed over the hills and valleys of Southern California, Don Mariano Reyes Roldan was granted 6,698 acres by the Mexican government and named his land Rancho Cañada de La Habra. The year was 1839, and the name referred to the "pass through the hills," the natural path to the north that was first documented by Spanish explorers in 1769 and was once the homeland of the Tongva people. In the 1860s, Abel Stearns, an American trader from Massachusetts, purchased Rancho Cañada de La Habra. Soon thereafter, heavy flooding, followed by a severe drought, brought bankruptcy to many cattle ranchers.

New settlers arrived to buy parcels of land on Rancho Cañada de La Habra in the 1890s. Most came by rail, others by wagon. These newcomers reaped a golden harvest of grains from the lands and raised sheep. Among the early settlers were the Milhous family, grandparents of the nation's 37th president, Richard Milhous Nixon. Nixon opened his first law office in La Habra adjacent to the civic center. The community was founded in 1896 and named "La Habra" when a US post office was started in a corner store off present-day La Habra Boulevard and Euclid Street, an intersection that continues to be vital to the city's civic life. A rural two-story schoolhouse was built the same year to accommodate the new influx of residents, a project spearheaded by one of La Habra's founding fathers, Willets J. Hole, a land developer who owned many large tracts in the area.

After many early struggles with droughts, irrigation lines brought precious water from the San Gabriel River area, allowing farmers to cultivate the semiarid land. First the walnut and then the citrus and avocado industries flourished here with the help of the new infrastructure. In 1908, the Pacific Electric Railroad line was completed through La Habra. In 1912, the Standard Oil Company established the Coyote Hills District and built a pipeline to El Segundo the same year. In 1913, La Habra became what was known as a "tent city" due to an influx of oil and citrus workers. By 1916, the area boasted a bank, three general stores, a meat market, a small hotel, two restaurants, a barbershop, a bakery, a dance hall, and a local newspaper. Industry was in full production, and property values skyrocketed.

The city was incorporated under general law on January 20, 1925, with a population of 3,000. By the time the police force was organized a year later in 1926, it employed only a chief, a traffic officer, and a patrolman. By 1928, influenced by the new Hass avocado grown nearby, the city bore the distinction of being the largest avocado center in Southern California. In 1930, the first fire department building was constructed, followed by the original city hall in 1935. By 1950, the population reached nearly 5,000, but the city experienced an even larger boom within that decade to over 25,000 residents. The civic center took shape when the existing county library was dedicated in 1966, and a second administration building opened in 1969. In 2017, this building was replaced by the current city hall, just across La Habra Boulevard from its predecessor, in order to better serve the needs of our present-day citizens.

In 1974, the National Municipal League honored La Habra with the prestigious All-America City award. The organization found La Habra to be outstanding in the areas of human relations, housing needs, environmental protection, business revitalization, broadening educational opportunities, boosting industry, and community services. The city was also recognized for its contributions toward improving the effectiveness of civic organizations. At the time, over 100 service clubs and civic organizations contributed to the city's progress and success, and this integral community participation continues today. La Habra was also chosen as a Bicentennial City during 1974 in recognition of its commemoration of this significant year in American history.

The city, through council policy and the leadership of its management team, has maintained its reputation for finding innovative solutions to local needs. Through the years, numerous programs have been implemented, including the formation of a neighborhood housing service, a local development company (now known as Landmark Development Co.), and a child development program. The city and community supporters also built the first children's museum west of the Mississippi River. The police department continues to administer model programs for crime prevention and community-based policing.

Located at Orange County's northernmost corner, La Habra today is 7.3 square miles with a population of nearly 62,000. A quiet bedroom community, it is conveniently located within an hour's drive of many beaches, mountains, and desert recreation areas, as well as the cultural center of Los Angeles.

A full-service city with over 250 full-time employees, La Habra provides a full range of services, including police and fire protection, water and sanitation services, street maintenance, and animal control, to name a few. Police equipment, as well as the communications system, is state-of-the-art. The city's water supply has been diversified to produce a lower-cost and more flexible delivery system for the residents. Within the city are three outstanding school districts that, combined, operate eight elementary schools, three middle schools, and two high schools. A number of private and specialty schools also call La Habra home, including preschools and those for the developmentally challenged. The city continues to focus on public safety, economic development, beautification, and infrastructure improvements.

La Habra is committed to providing recreational, cultural, educational, health and wellness, and social services programs and special events for the community. The city of La Habra offers 22 parks, a children's museum, community theater, tennis center, child development program, employment and training youth program, and other affordable and accessible programs for youth, families, active adults, and seniors. While the motto "A Caring Community" was only adopted within the last few decades, it remains an apt description for the city and for La Habrans from the area's humble beginning to today.

One

Early Days in the La Habra Valley

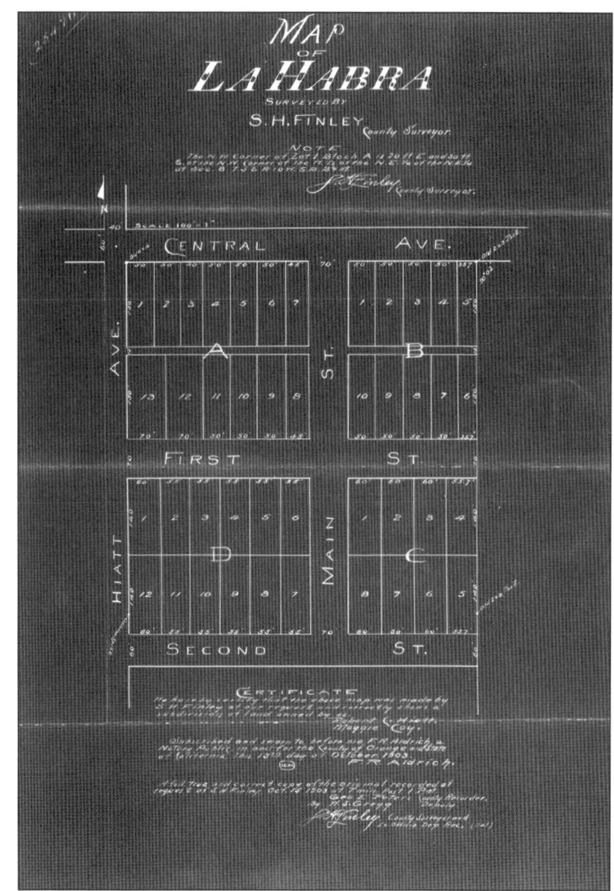

This map shows the cartography of a section of La Habra owned by Robert Hiatt in 1903. The self-named Hiatt Avenue became the first north-south street in the town to be paved in 1915; La Habra was gaining a reputation for having more paved roads in Orange County than any other community of its size. The road has been known as Euclid Street since 1961. (Courtesy of Elana Samson.)

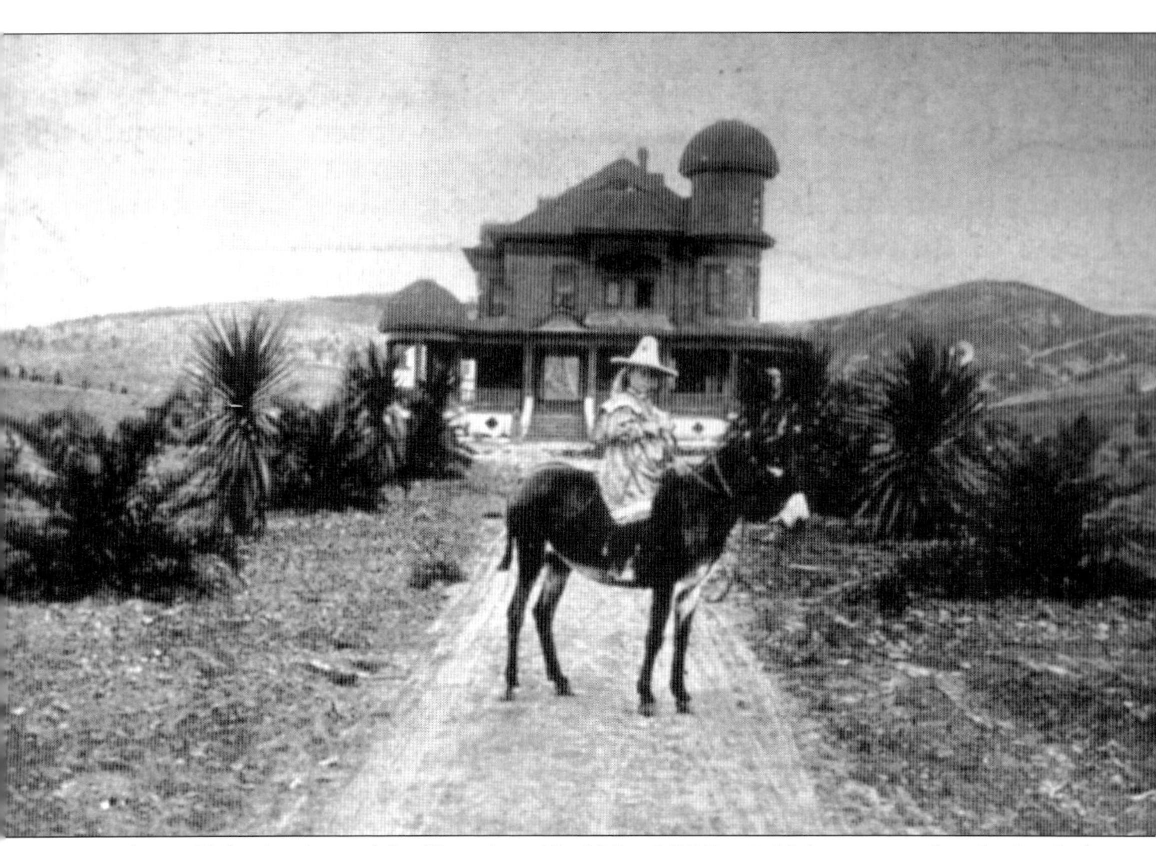

Agnes Hole, daughter of the "founder of La Habra," Willets J. Hole, sits atop her donkey Judge, whom she often used as a mode of transportation to the La Habra School. Agnes served as part of the inspiration for her father's push to form a school district in La Habra, though he also knew that selling land to new settlers would be easier if there was a school in the valley. Seen behind her is Hole mansion, which was located north of Whittier Boulevard between Walnut and Citrus Streets. Hole mansion was built with bricks made from La Habra clay that were manufactured directly on the property. During the construction, the Hole family lived in the carriage house, just off-screen of this photograph. The $10,000 mansion was completed in 1896 and was noted as one of the most distinctive homes in Southern California. (Courtesy of the La Habra Historical Society.)

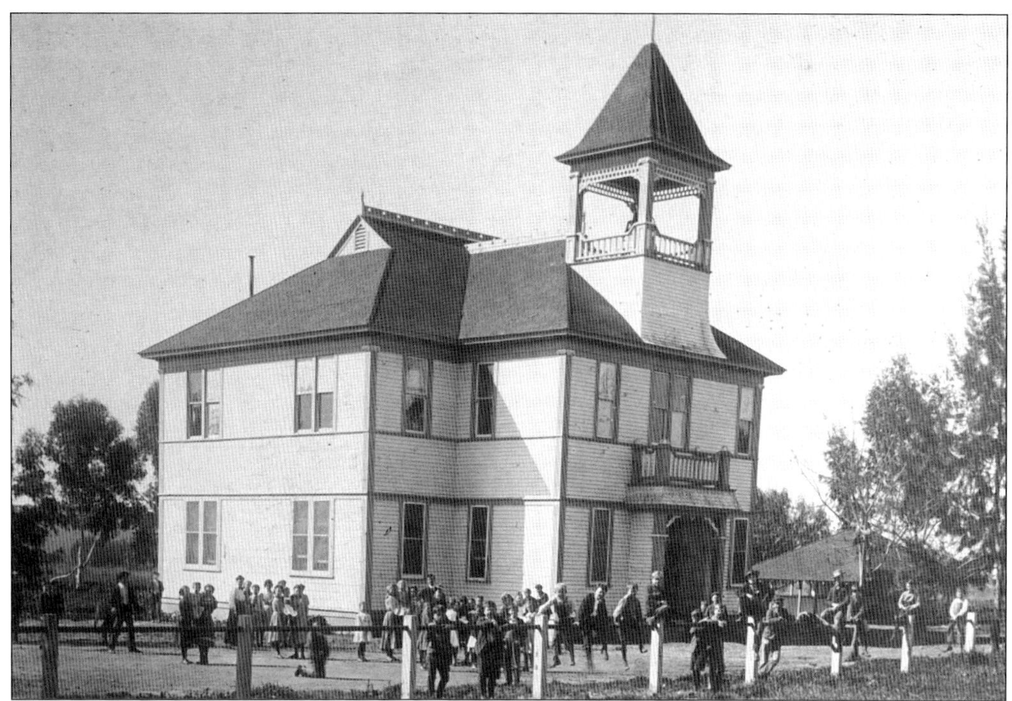

La Habra School became the very first school in the community when it was built in 1896 on Central Avenue. The schoolhouse hosted many classes and community events until it was turned into apartment homes following the construction of the new La Habra Grammar School. A group of students and staff, some posing and some unaware of the camera, can be seen on the lawn of the La Habra School in 1903. The baseball team of 1912 is also seen posing proudly with their equipment on the steps of La Habra School nearly a decade later. (Above, courtesy of the La Habra Historical Society; below, courtesy of the Orange County Archives.)

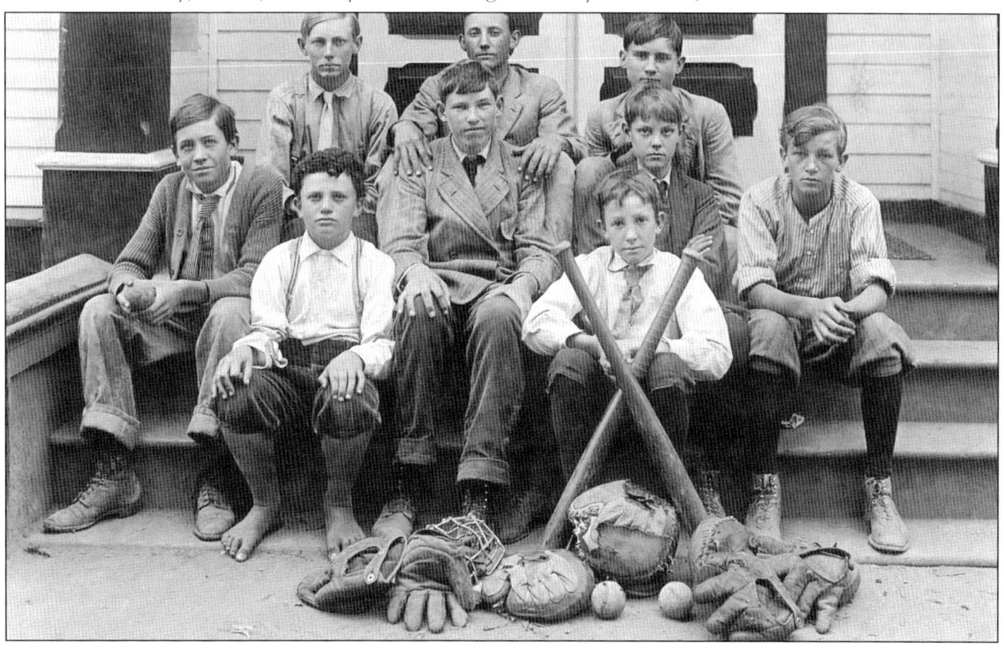

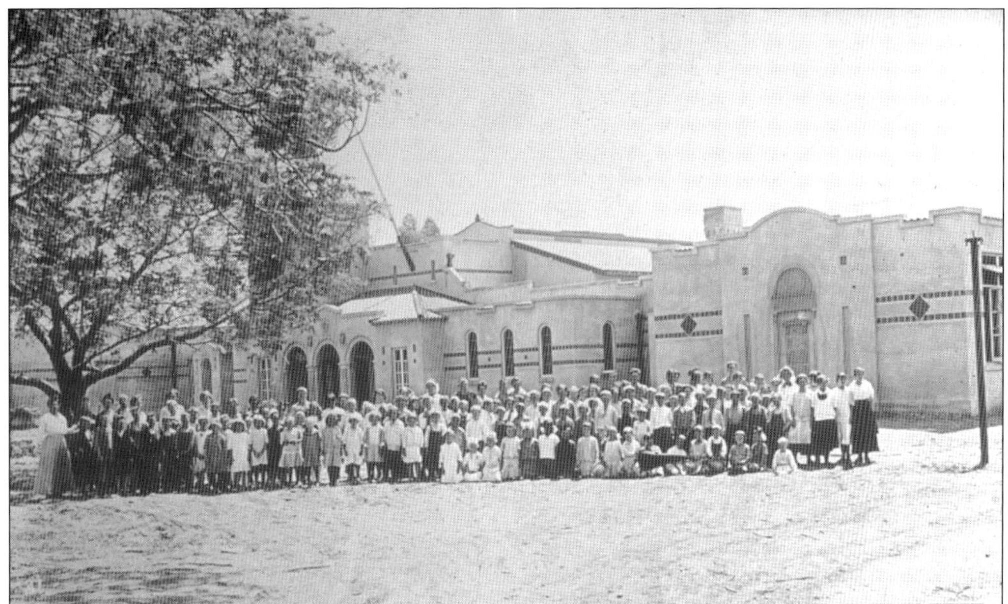

Faced with overcrowding, the new La Habra Grammar School building replaced the old schoolhouse in 1915. This group portrait in front of the new La Habra Grammar School, located on Central Avenue, was taken in the school's inaugural year. It was renamed Washington School in 1923 and currently operates as a middle school. (Courtesy of the La Habra Historical Society.)

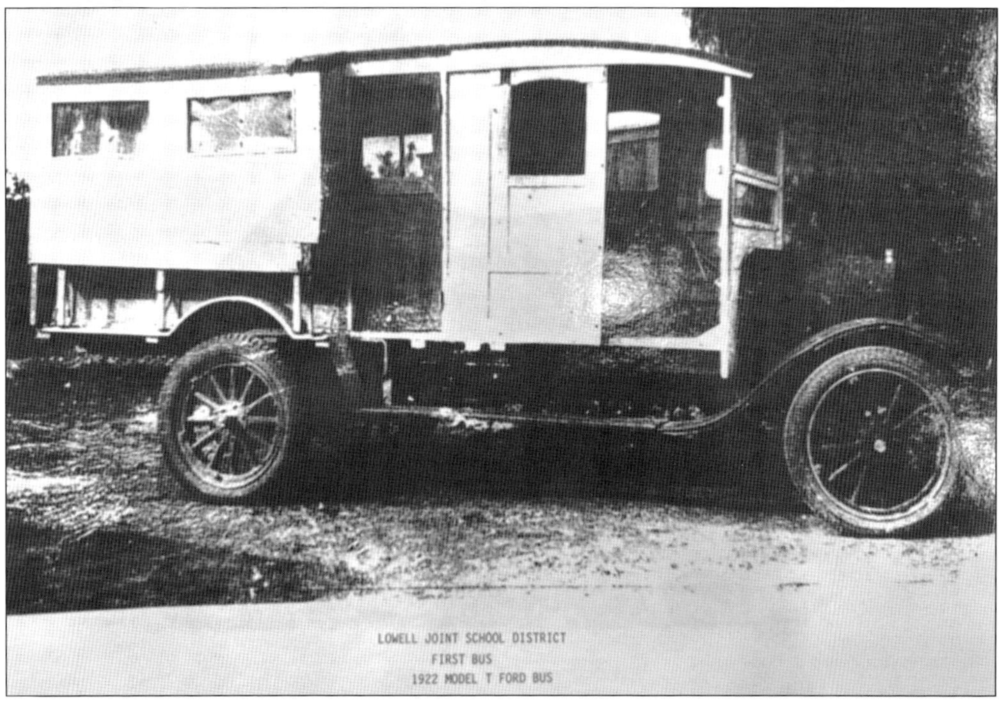

The Lowell Joint School District acquired its first motorized school bus in 1922 with this Ford Model T bus, which was cutting-edge technology for the time. Before this advancement, the area utilized four horse-drawn buses to transport students to and from class. (Courtesy of the City of La Habra.)

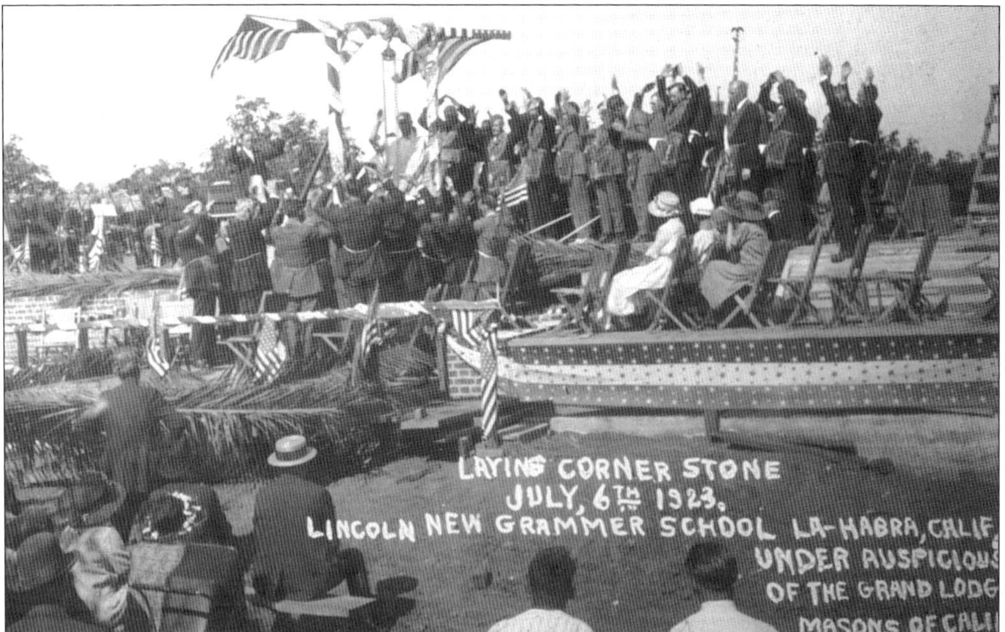

The cornerstone ceremony for Lincoln School on July 6, 1923, was sponsored by the Grand Masonic Lodge of California, and a copper box containing historical items was placed within the cornerstone. After Lincoln School was built, La Habra Grammar School was renamed Washington School, and West Side School was renamed Wilson School to continue the presidential theme. (Courtesy of the La Habra Historical Society.)

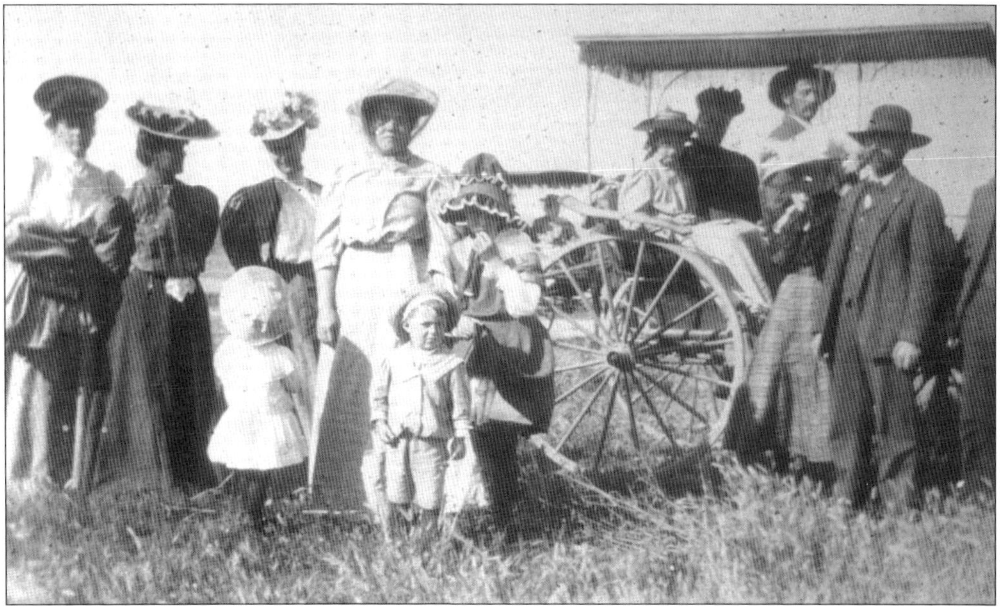

Early settler families, including those of soon-to-become prominent community members John Luehm and Jacob Leutwiler, can be seen here on an outing to the La Habra Valley. These families were among a group of settlers who moved from Highland, Illinois, to the La Habra area in the 1890s, the same time as this photograph was captured. (Courtesy of the La Habra Historical Society.)

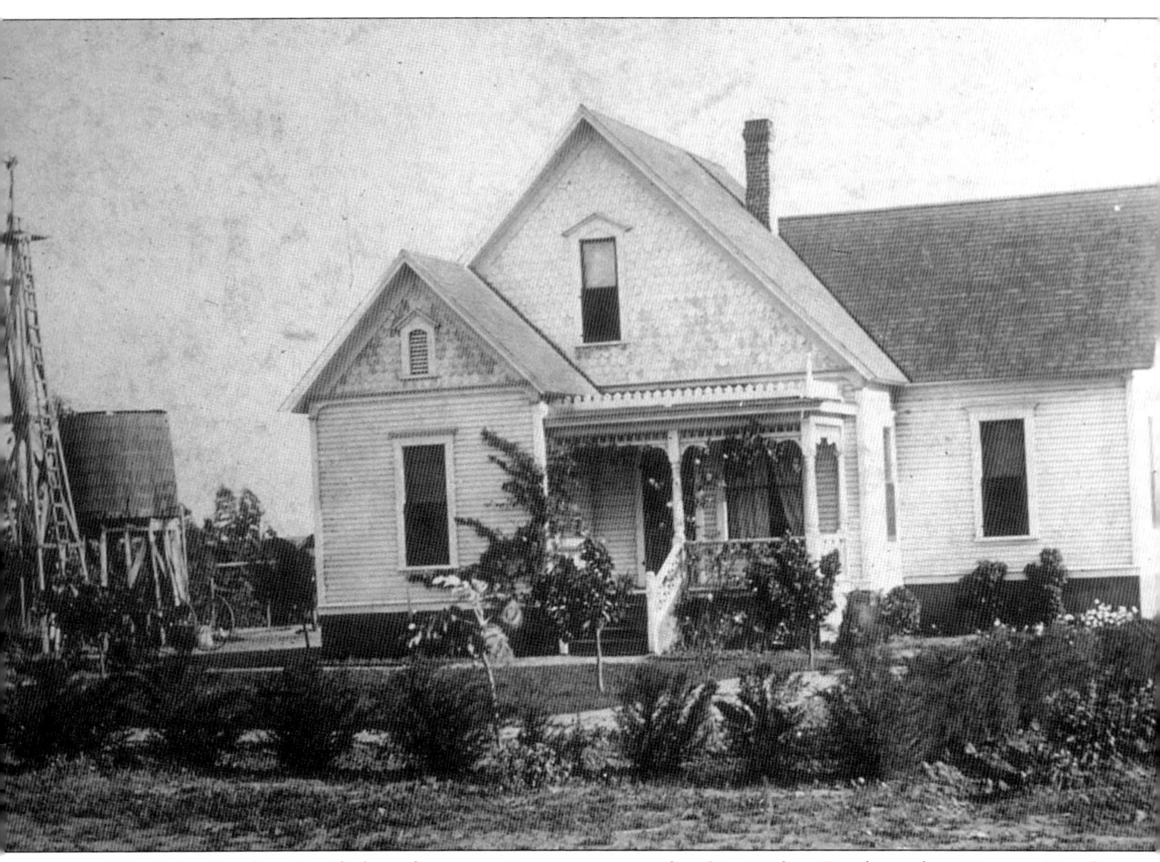

The John Luehm family bought extensive acreage south of La Habra Boulevard to Coyote Creek and was the first of this group of Swiss settlers from Illinois to make La Habra their home in 1897. This home, along with those of other early settlers, was featured in a community promotional brochure around 1905 to advertise the "most fertile and productive valley in the state." Luehm's property did prove to be on fertile and productive land, as he operated a farm that produced mostly fruit. He was later named the first vice president of the Farmers' Club, which was active in promoting agriculture in the La Habra Valley and was a predecessor to the local chamber of commerce. (Courtesy of the La Habra Historical Society.)

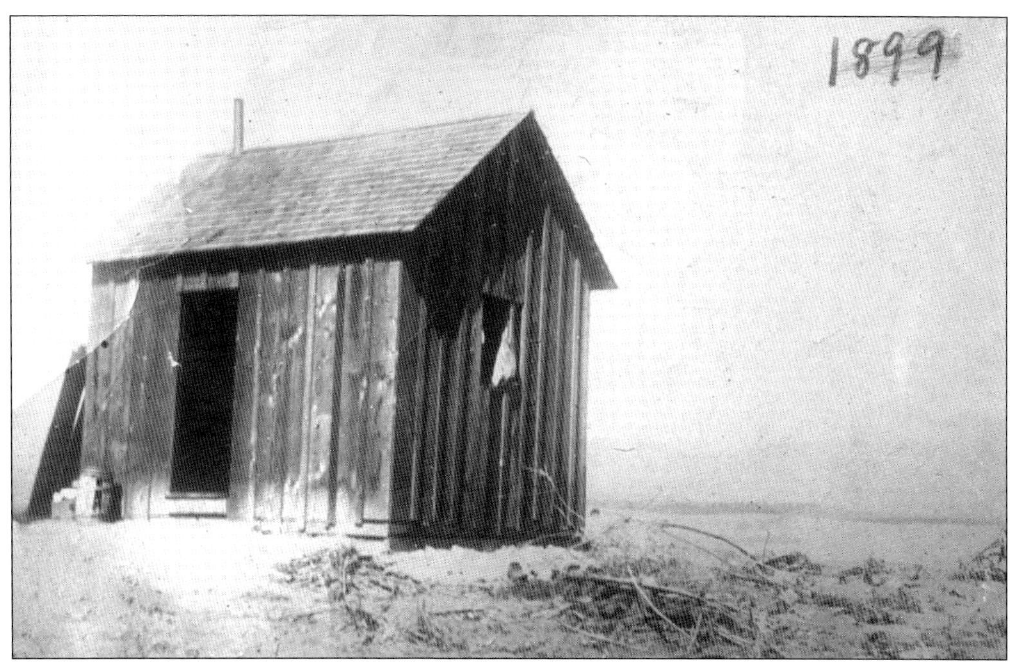

Deciding to settle down in La Habra after his successful survey trip, Jacob Leutwiler built his first small house, seen above in 1899, near the northwest corner of what is now Walnut Street and Lambert Road. The road in front of the house was named Ocean Avenue by his wife, as she believed the street would someday lead to the sea, but the street name was later changed to Lambert Road in the 1960s. Their second home, seen below, stood at the southeast corner of First Avenue and Euclid Street and was later used as John Frazier's Plumbing and a Mexican deli. (Both, courtesy of the La Habra Historical Society.)

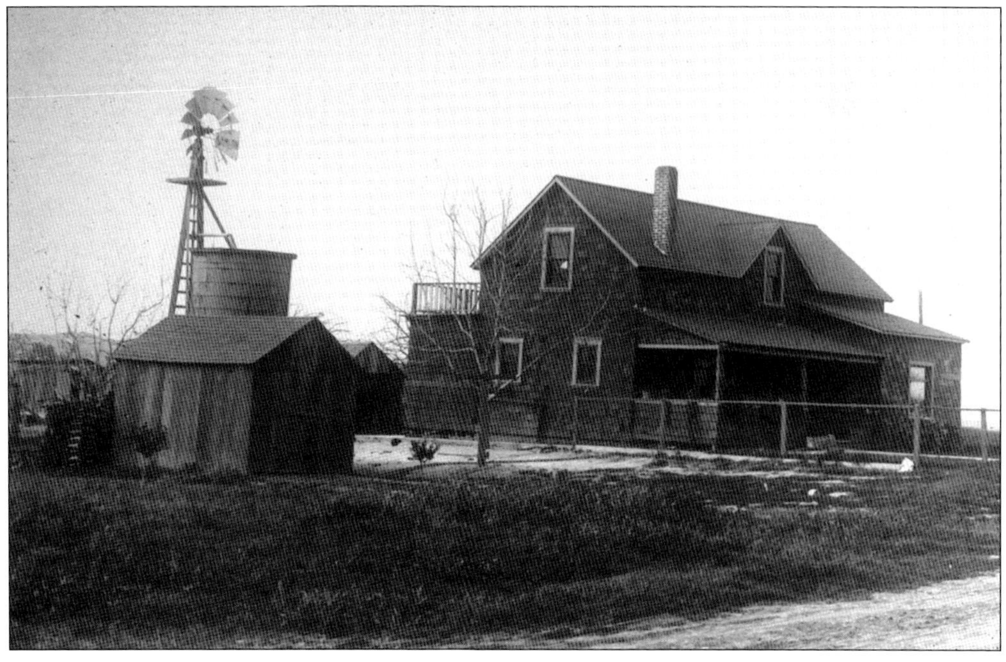

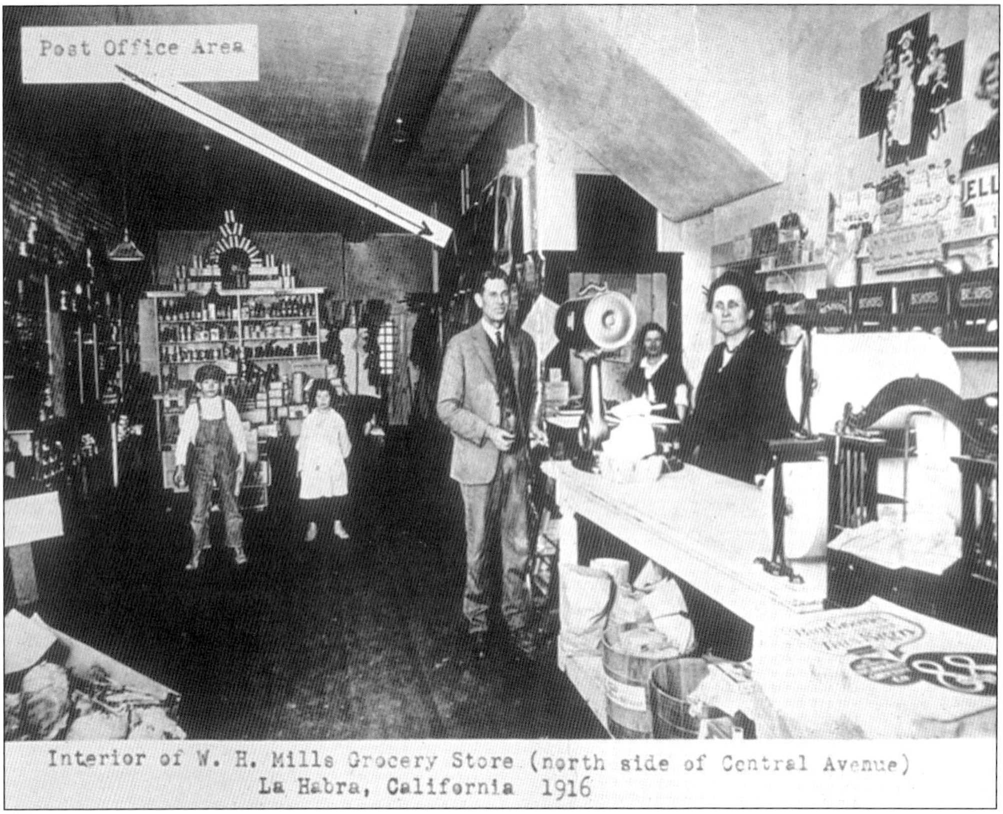

Post Office Area

Interior of W. H. Mills Grocery Store (north side of Central Avenue)
La Habra, California 1916

Located off of Central Avenue, the Coy's Store opened as La Habra's general store in 1895, run by postmaster Zachary Coy, brother-in-law of city founder Willets J. Hole. The first post office was added inside the store, officially naming the settlement "La Habra" and earning the intersection of current-day Euclid Street and La Habra Boulevard the nickname "the birthplace of La Habra." The store was renamed W.H. Mills Grocery after Coy's death; the Millses themselves can be seen at the counter along with assistant Katie Terwiliger in 1916. The progression of the store is seen a few years later in 1919. (Both, courtesy of the La Habra Historical Society.)

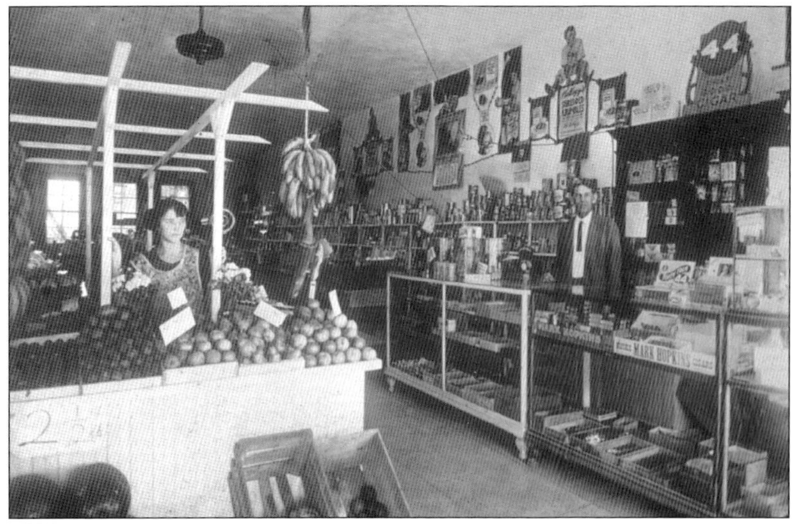

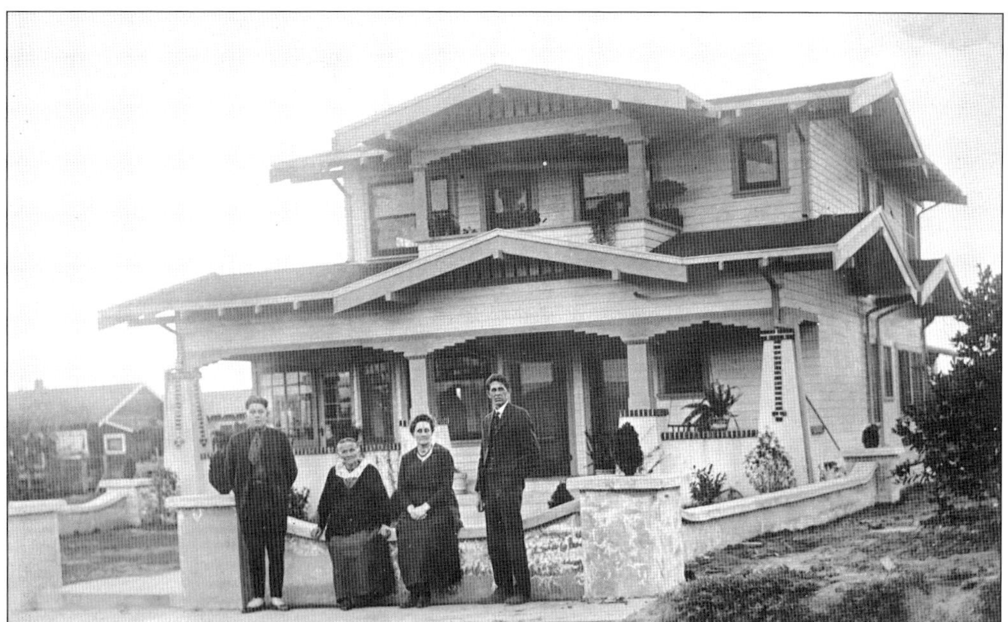

The Mills family of grocery store fame built this large home on North Hiatt Avenue, now Euclid Street, before World War I. It then became the Coleman Mortuary before it was sold to the City of La Habra, and it is now used as one of the facilities for the city's child development program. (Courtesy of the City of La Habra.)

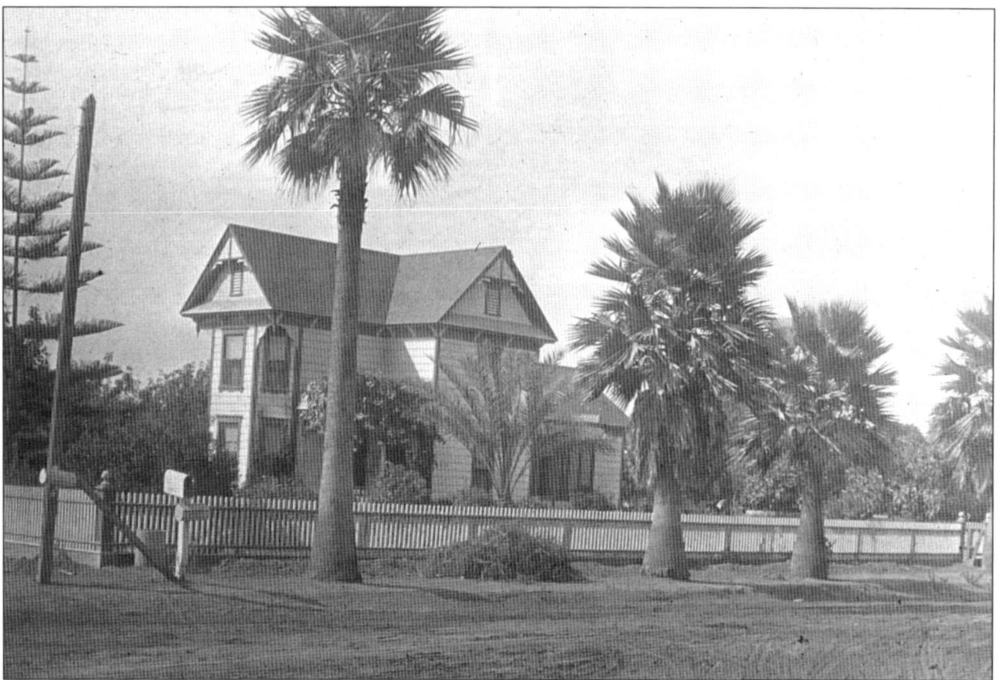

Prominent walnut farmer S.M. Smith built this two-story house on the southwest corner of modern-day La Habra and Beach Boulevards in 1895. His 40-acre farm was located on this property, which was eventually referred to as "Smith Corner" due to several Smith families residing on or near this corner at the time. (Courtesy of the La Habra Historical Society.)

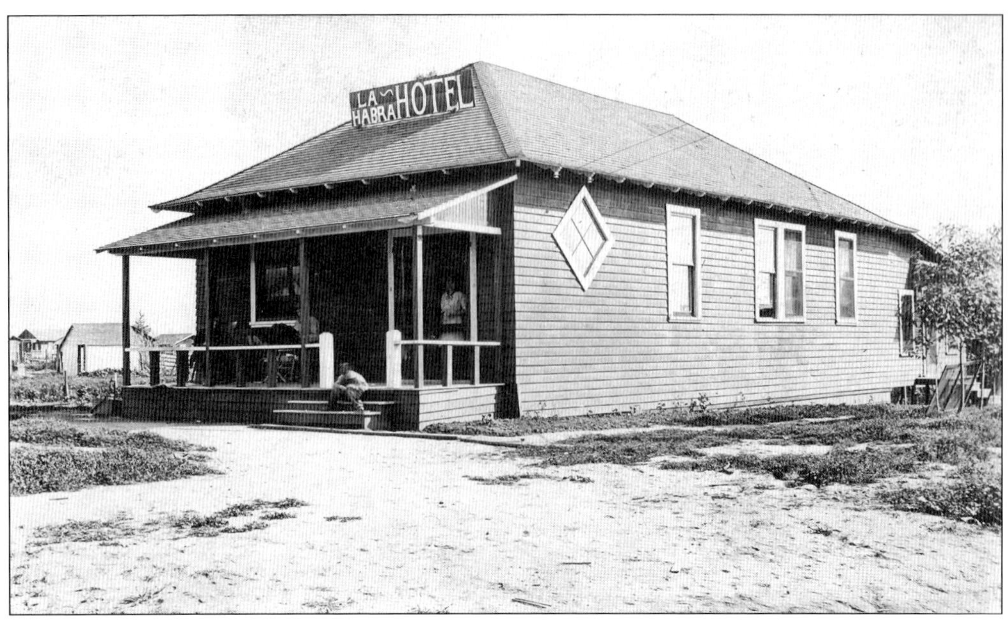

Shortly after moving to town in 1909, Mrs. Hersey's Boarding House opened to accommodate those visiting La Habra. By 1913, the boardinghouse had grown to such proportions that F.A. Hersey joined his wife as a business partner, expanded the boardinghouse to a larger location, and changed the name to the La Habra Hotel. The La Habra Hotel opened in the old Methodist Episcopal church location at Second Avenue and Main Street, just down the road from the church's new location on First Avenue and Main Street. Today, both buildings maintain the architecture seen above in 1914. (Above, courtesy of the Orange County Library; below, courtesy of the La Habra Historical Society.)

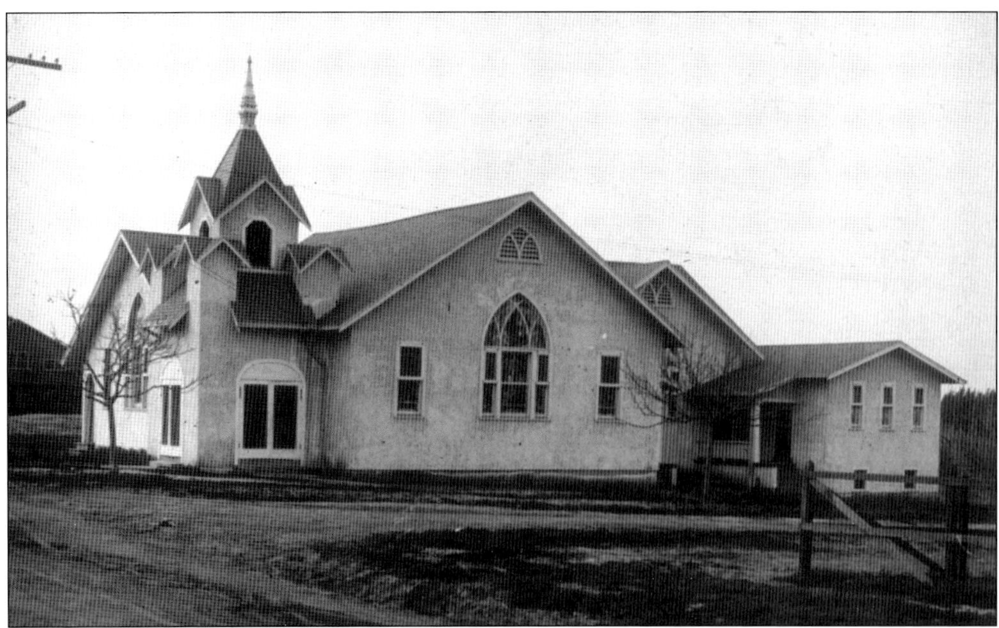

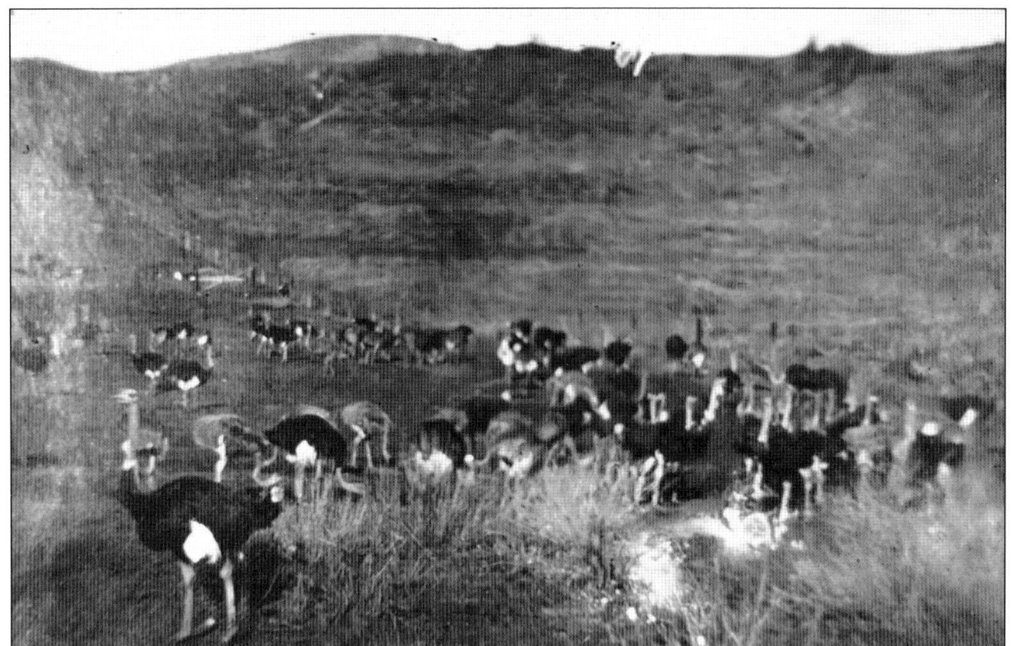

The Cawston Ostrich Farm was a unique La Habra attraction in the early 1900s. The farm, which was one of 10 Cawston Ostrich Farm locations, was housed on 200 acres of land east of Citrus Drive. The farm's La Habra location operated from 1904 to 1909 and attracted visitors to the city from far and wide; all Cawston Ostrich Farm locations ceased operations by 1935. (Courtesy of the La Habra Historical Society.)

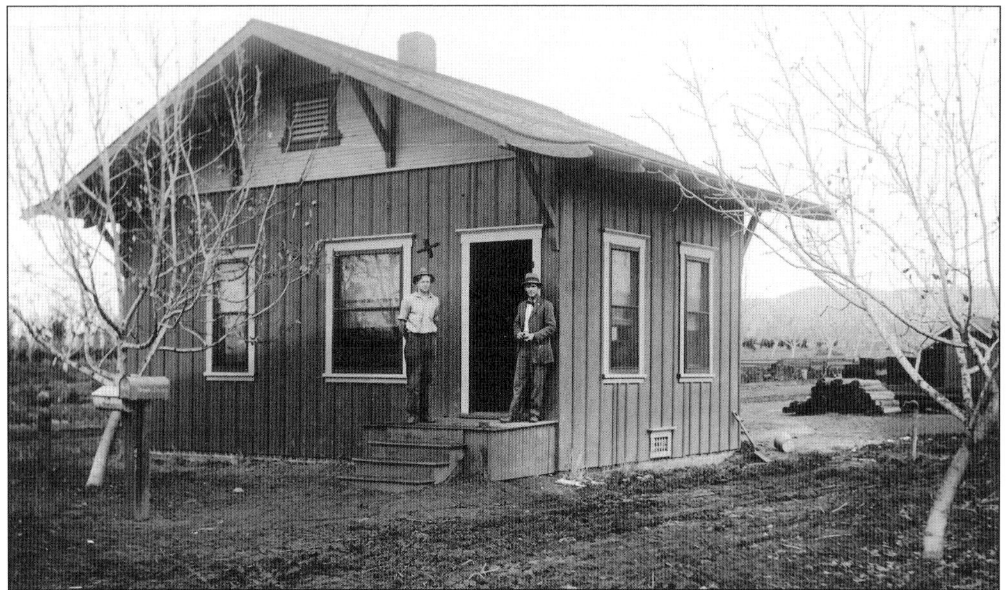

Opening in 1907 at the corner of Cypress Street and Central Avenue, the Brown & Dauser Lumber Yard proved to be a valuable business to La Habra when it provided much of the building materials needed during the development boom of the 1920s. Shown above in 1910, it later made headlines in January 1944 as the "oldest business in La Habra" when it first changed ownership. (Courtesy of the Thomas Pulley collection.)

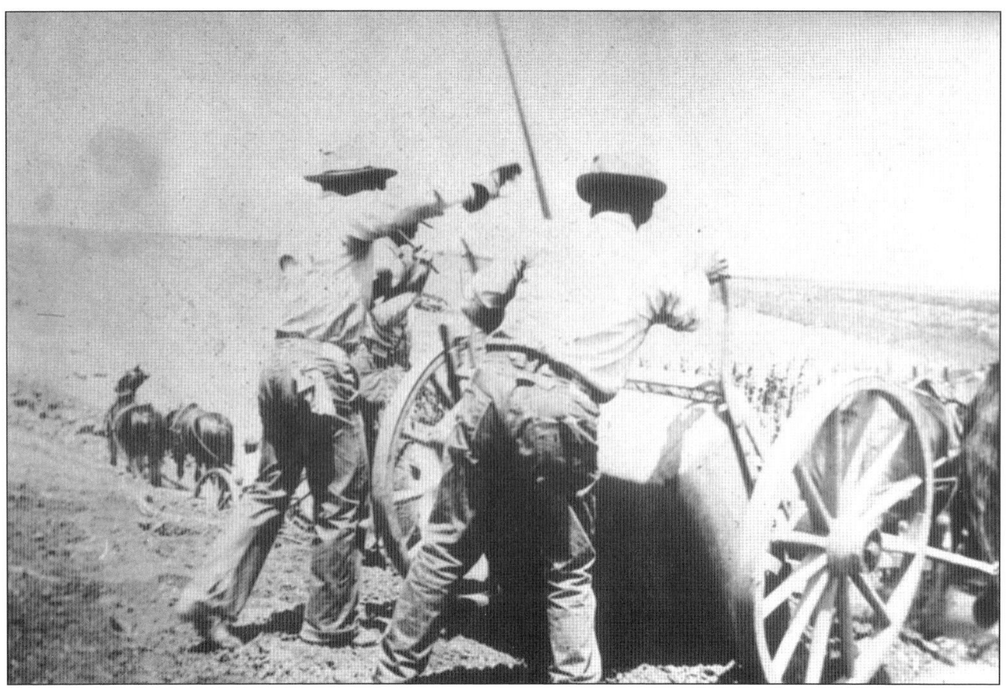

The extension of the Pacific Electric line from Whittier to La Habra was due largely to W.J. Hole's efforts to bring the first railroad to this area to meet the transportation needs of the residents, as well as the citrus and oil industries in the valley. Several farmers fought against the efforts as they did not want the route to go through their land and refused to sign the right-of-way contracts necessary for development, while other La Habra residents fundraised to repay landowners for any damages resulting from the proposed railroad to encourage their permission. The construction of the line from Whittier to La Habra started in 1907. (Both, courtesy of the La Habra Historical Society.)

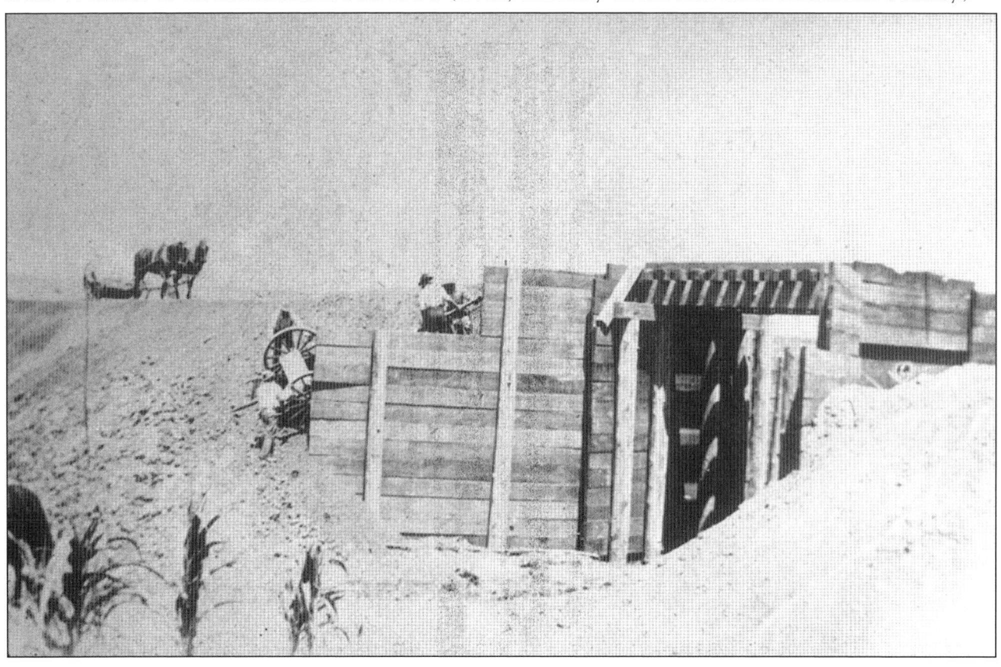

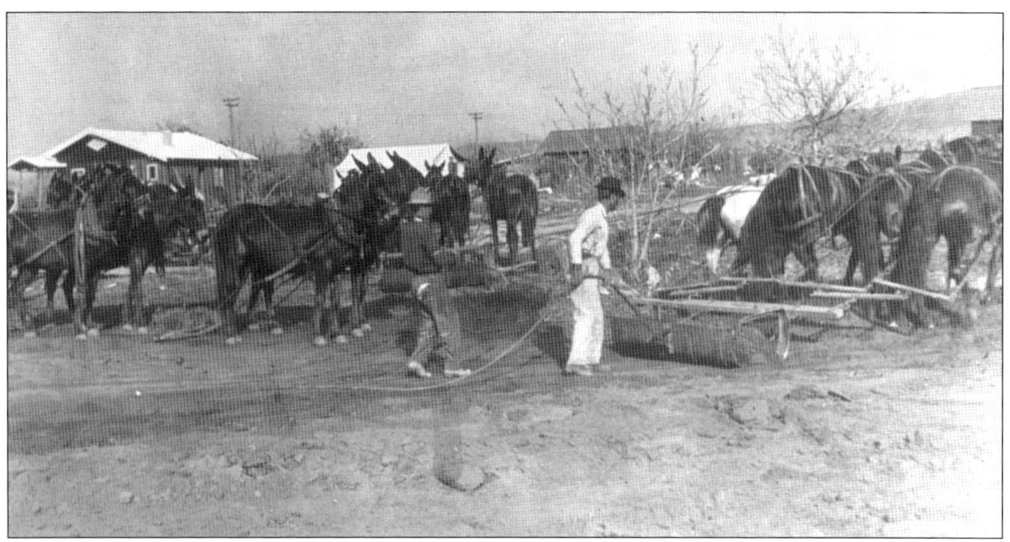

Road crews of the 1910s, such as this one of La Habra, had little technology available to help with the maintenance of the streets. Their work instead primarily consisted of leveling bumps and removing obstacles. They are seen here using livestock to drag machinery for leveling the street currently known as La Habra Boulevard. (Courtesy of the La Habra Historical Society.)

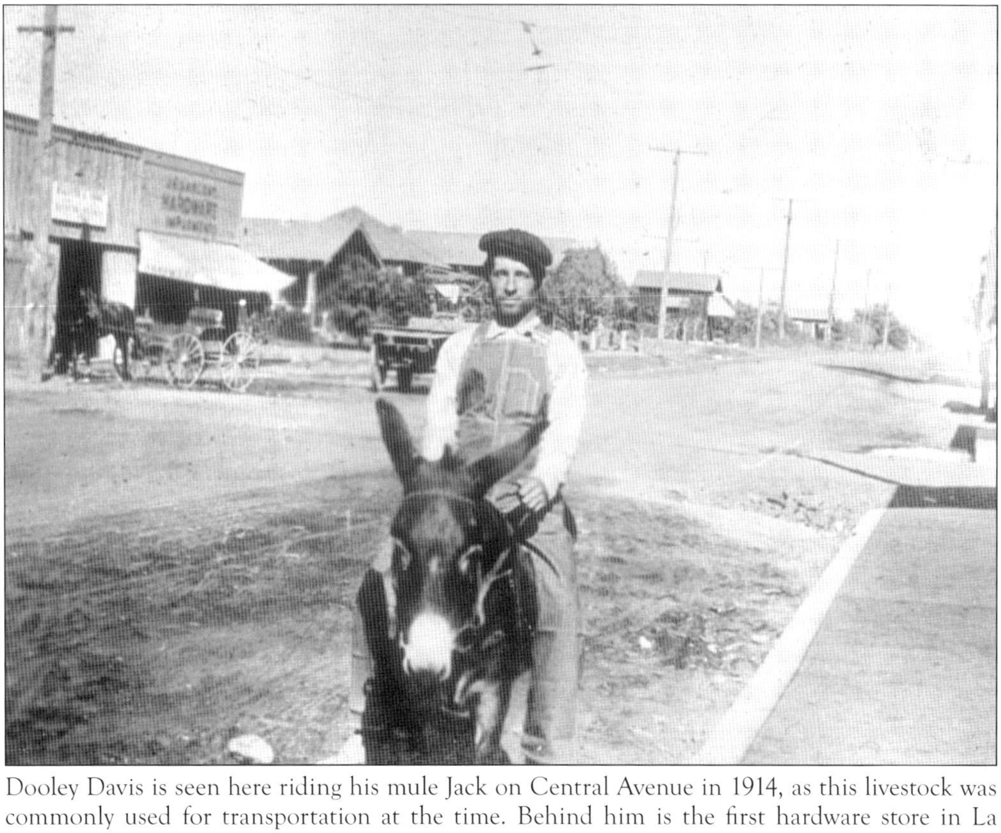

Dooley Davis is seen here riding his mule Jack on Central Avenue in 1914, as this livestock was commonly used for transportation at the time. Behind him is the first hardware store in La Habra, which opened after local businessman J.C. Sargant used profits from selling his blacksmith shop to open the hardware store that same year. (Courtesy of the La Habra Historical Society.)

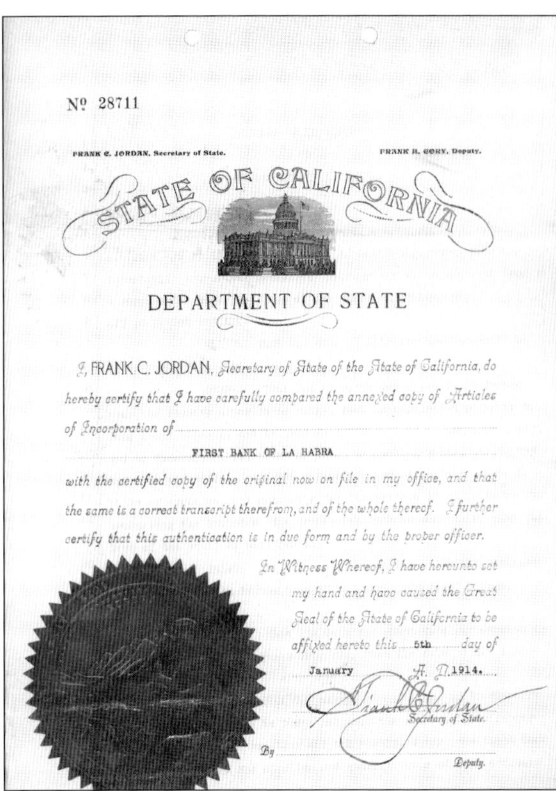

No 28711

FRANK C. JORDAN, Secretary of State. FRANK H. CORY, Deputy.

STATE OF CALIFORNIA

DEPARTMENT OF STATE

I, FRANK C. JORDAN, Secretary of State of the State of California, do hereby certify that I have carefully compared the annexed copy of Articles of Incorporation of FIRST BANK OF LA HABRA with the certified copy of the original now on file in my office, and that the same is a correct transcript therefrom, and of the whole thereof. I further certify that this authentication is in due form and by the proper officer.

In Witness Whereof, I have hereunto set my hand and have caused the Great Seal of the State of California to be affixed hereto this 5th day of January A. D. 1914.

Frank C. Jordan
Secretary of State.

By .... Deputy.

This document certifies the incorporation of the First Bank of La Habra by California's secretary of state in 1914. When the bank opened its doors, it had $25,000 in capital, which, adjusted for inflation, would be worth almost $790,000 in 2025. (Courtesy of Elana Samson.)

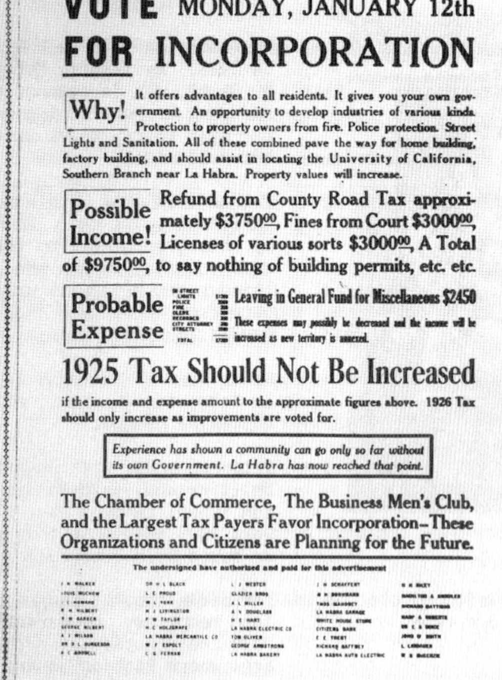

VOTE MONDAY, JANUARY 12th

FOR INCORPORATION

Why! It offers advantages to all residents. It gives you your own government. An opportunity to develop industries of various kinds. Protection to property owners from fire. Police protection. Street Lights and Sanitation. All of these combined pave the way for home building, factory building, and should assist in locating the University of California, Southern Branch near La Habra. Property values will increase.

Possible Income! Refund from County Road Tax approximately $3750.00, Fines from Court $3000.00, Licenses of various sorts $3000.00, A Total of $9750.00, to say nothing of building permits, etc. etc.

Probable Expense Leaving in General Fund for Miscellaneous $2450. These expenses may possibly be decreased and the income will be increased as new territory is annexed.

1925 Tax Should Not Be Increased

if the income and expense amount to the approximate figures above. 1926 Tax should only increase as improvements are voted for.

Experience has shown a community can go only so far without its own Government. La Habra has now reached that point.

The Chamber of Commerce, The Business Men's Club, and the Largest Tax Payers Favor Incorporation—These Organizations and Citizens are Planning for the Future.

The undersigned have authorized and paid for this advertisement

Those in favor of incorporating La Habra heavily advertised the incorporation election, which was held on January 12, 1925. The ticket was successful with a 311-146 vote to incorporate, and the papers officially making La Habra a city arrived on the official incorporation date of January 20, 1925. (Courtesy of the La Habra Historical Society.)

Two

Laying a Strong Foundation

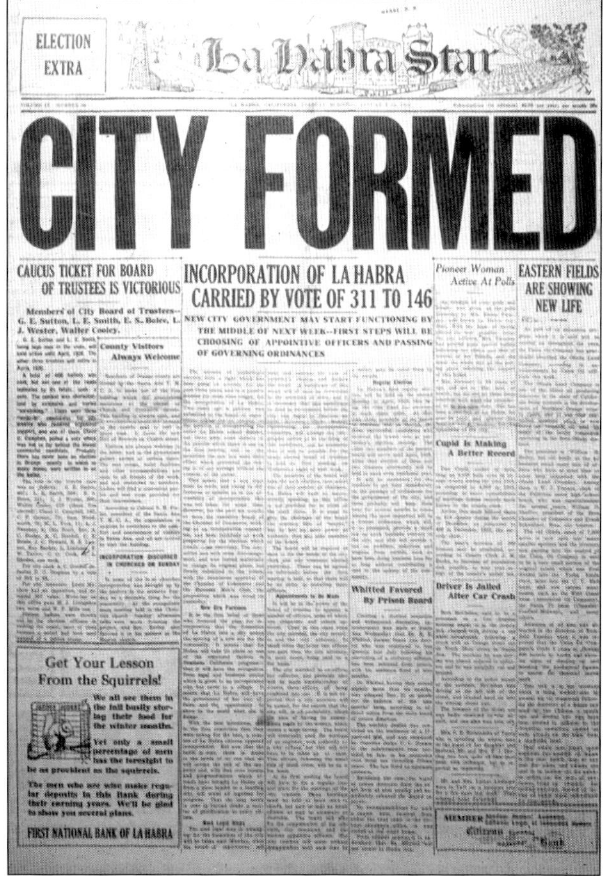

ELECTION EXTRA

La Habra Star

CITY FORMED

CAUCUS TICKET FOR BOARD OF TRUSTEES IS VICTORIOUS

Members of City Board of Trustees— G. E. Sutton, L. E. Smith, E. S. Bolce, L. J. Wester, Walter Cooley.

INCORPORATION OF LA HABRA CARRIED BY VOTE OF 311 TO 146

NEW CITY GOVERNMENT MAY START FUNCTIONING BY THE MIDDLE OF NEXT WEEK—FIRST STEPS WILL BE CHOOSING OF APPOINTIVE OFFICERS AND PASSING OF GOVERNING ORDINANCES

Pioneer Woman Active At Polls

EASTERN FIELDS ARE SHOWING NEW LIFE

County Visitors Always Welcome

Cupid Is Making A Better Record

Whitted Favored By Prison Board

Driver Is Jailed After Car Crash

Get Your Lesson From the Squirrels!

We all see them in the fall busily storing their food for the winter months.

Yet only a small percentage of men has the foresight to be as provident as the squirrels.

The men who are wise make regular deposits in this Bank during their earning years. We'll be glad to show you several plans.

FIRST NATIONAL BANK OF LA HABRA

The 311-146 vote to incorporate the City of La Habra was commemorated as front page news on the *La Habra Star* in January 1925. Stating that next steps included appointing officers and passing government ordinances as soon as the following week, this cover serves as a reminder of the collective efforts and shared vision that led to La Habra's cityhood from its founding in 1896. (Courtesy of the La Habra Historical Society.)

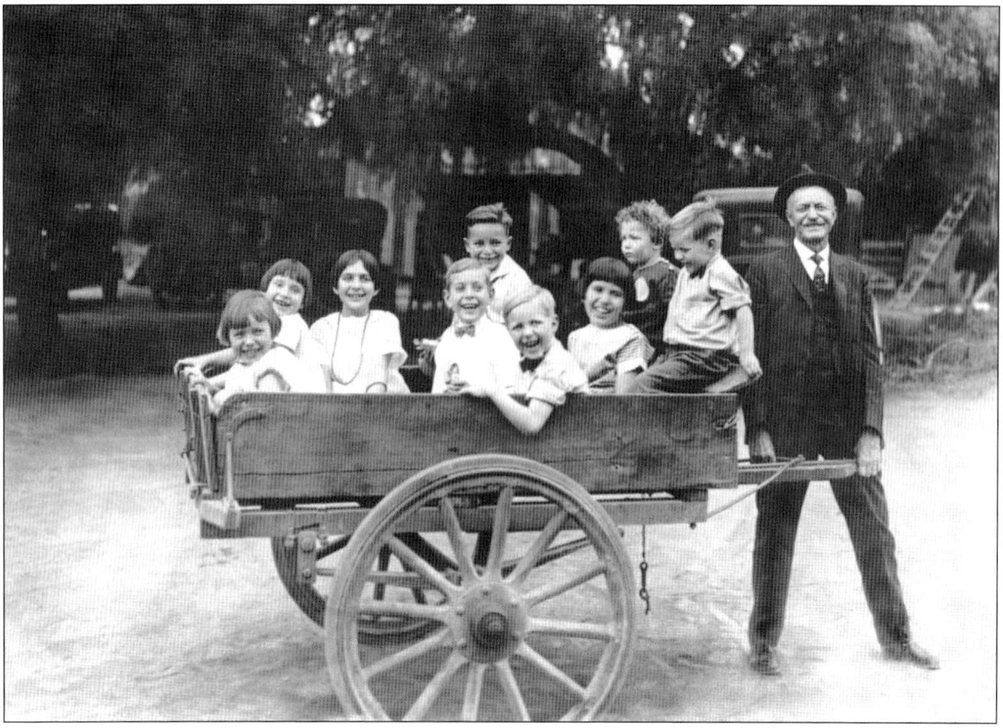

After the incorporation, John Launer was elected the first official mayor of La Habra. Here, he proudly poses with his grandchildren (from left to right) Jean, Ruth Marie, Kathryn, Leland, Malcolm, Raymond, Eunice, Douglas, and Jimmy in 1925. Launer was an active member of the community and a local farmer. This photograph reflects the spirit of community and family that has long been synonymous with La Habra. (Courtesy of the La Habra Historical Society.)

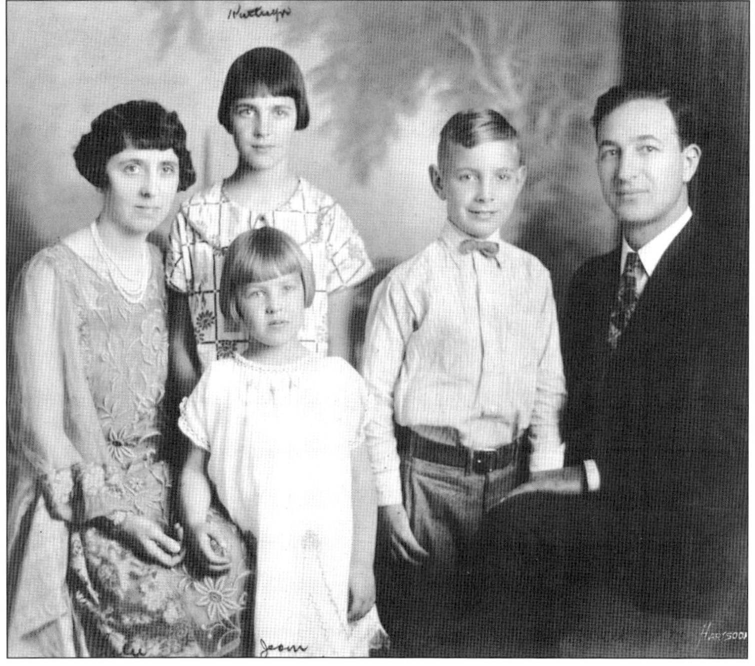

Albert Launer (right) poses with his wife, Lulu Convis (left), and children. One of the four sons of the first mayor, John Launer, he was the first city attorney, an active member of the Kiwanis Club, and, as a young lad, was a delivery boy for Zachary Coy's general store. (Courtesy of the Orange County Public Library.)

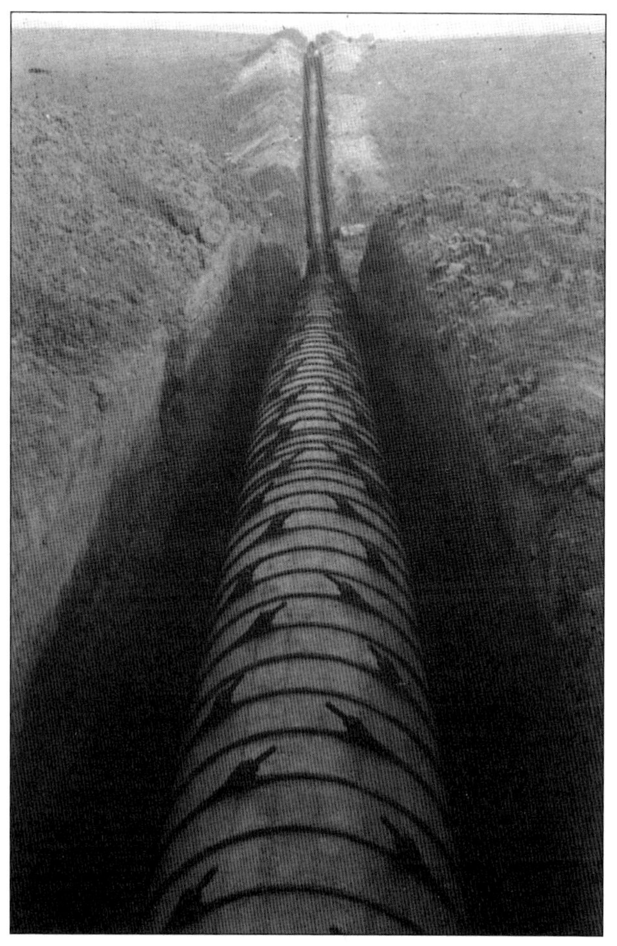

In 1906, a redwood pipeline was built to pump water from East Whittier to La Habra. The water originated in artesian wells on the San Gabriel River and then flowed to East Whittier by both natural and man-made channels. This photograph shows the pipeline freshly laid in 1906, providing water to both the community members and multiple farms of La Habra. To retrieve water from the pipe, pumphouses like this one in front of the J.S. Spotts house in La Habra held equipment needed to extract and distribute water to other locations, acting as a type of water supply control center. (Both, courtesy of the La Habra Historical Society.)

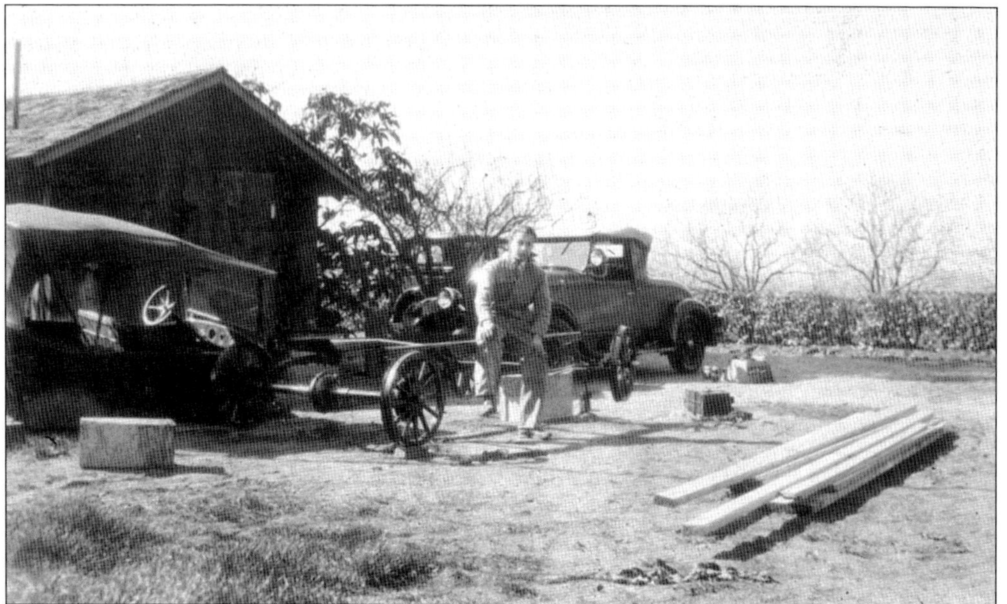

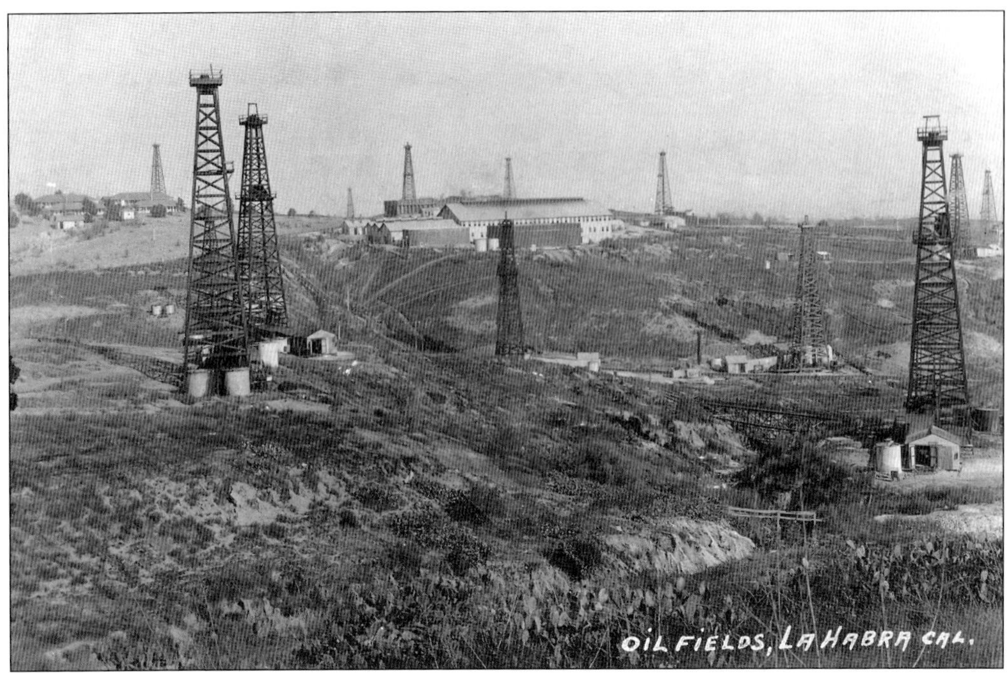

The discovery of oil in the region transformed La Habra's economic system from that of a small agricultural community into a thriving industrial center. California's agricultural markets were rapidly expanding, and the increased oil production helped to power the development of the industry. Pictured above is the front of a postcard that advertises La Habra's oil fields in 1915. With the oil boom swiftly expanding, cottages were constructed on and near the field sites to provide housing for workers and management. The photograph below shows a grouping of seven cottages housed at the oil fields of La Habra. (Both, courtesy of the Orange County Archives.)

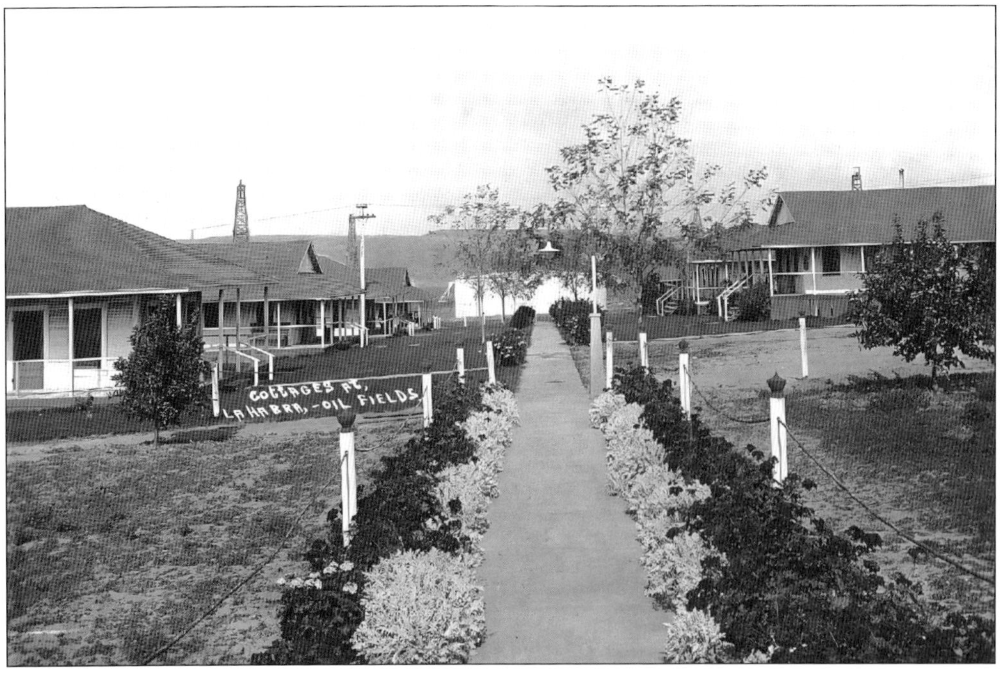

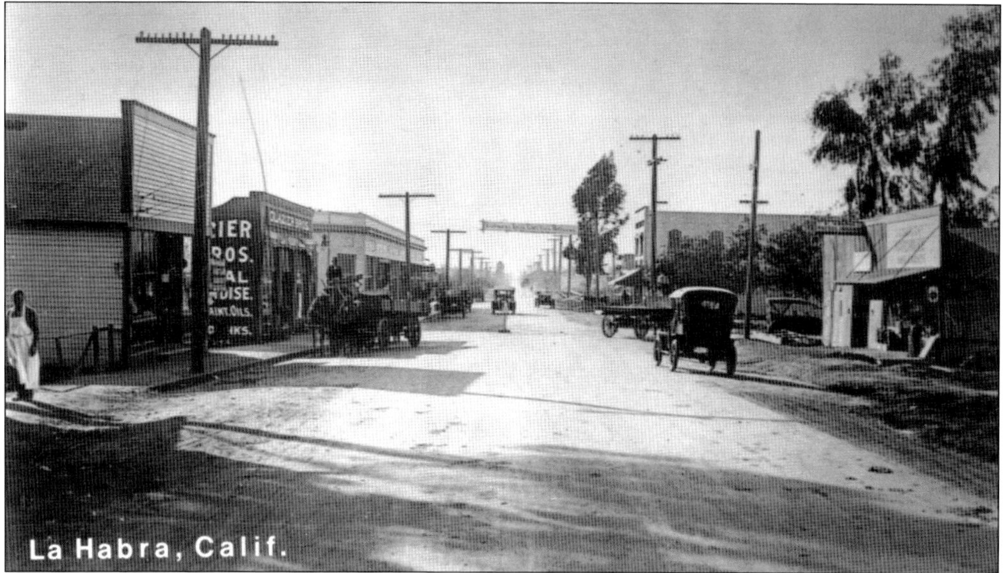

La Habra is shown as it appeared during World War I. The blacksmith shop of Eugene Young is on the right, and the Wester Hotel, where eccentric recluse Benjamin "Walking" Elliot lived, is the two-story brick building near the right-center of the photograph. Young and Elliot were famously involved in an armed altercation after Elliot refused to participate in La Habra's contribution to the war effort. (Courtesy of the Thomas Pulley collection.)

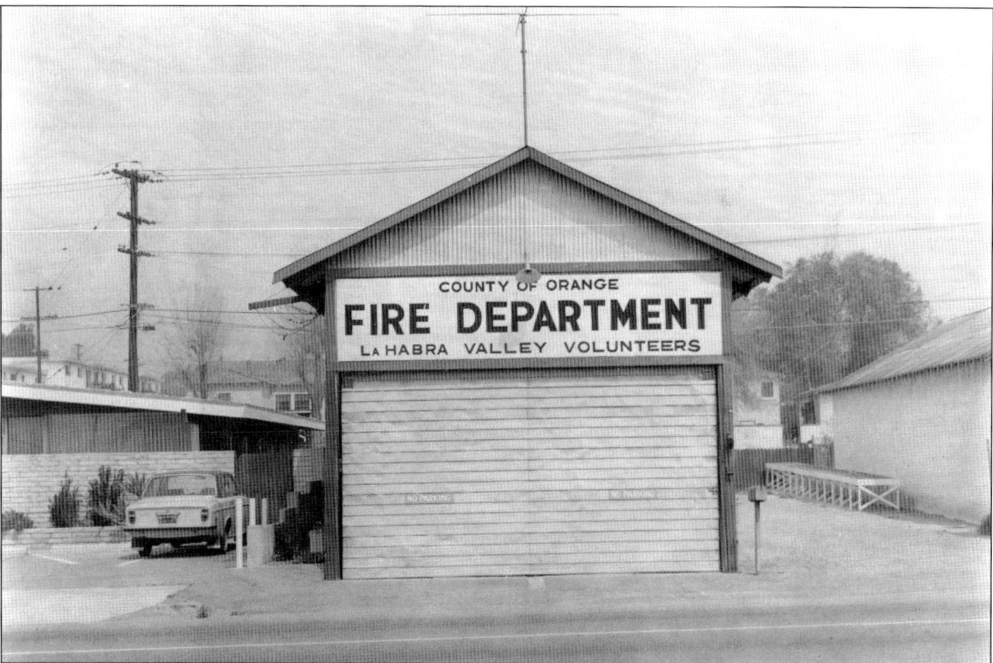

The La Habra Fire Department was founded in 1916 by a volunteer group of concerned citizens. The department was led by first "fire chief" Ralph Glazier and was largely funded by community fundraising. With La Habra's incorporation in 1925, equipment was handed over to the city, which provided services until the Los Angeles County Fire Department assumed fire protection service in 2005. (Courtesy of the La Habra Historical Society.)

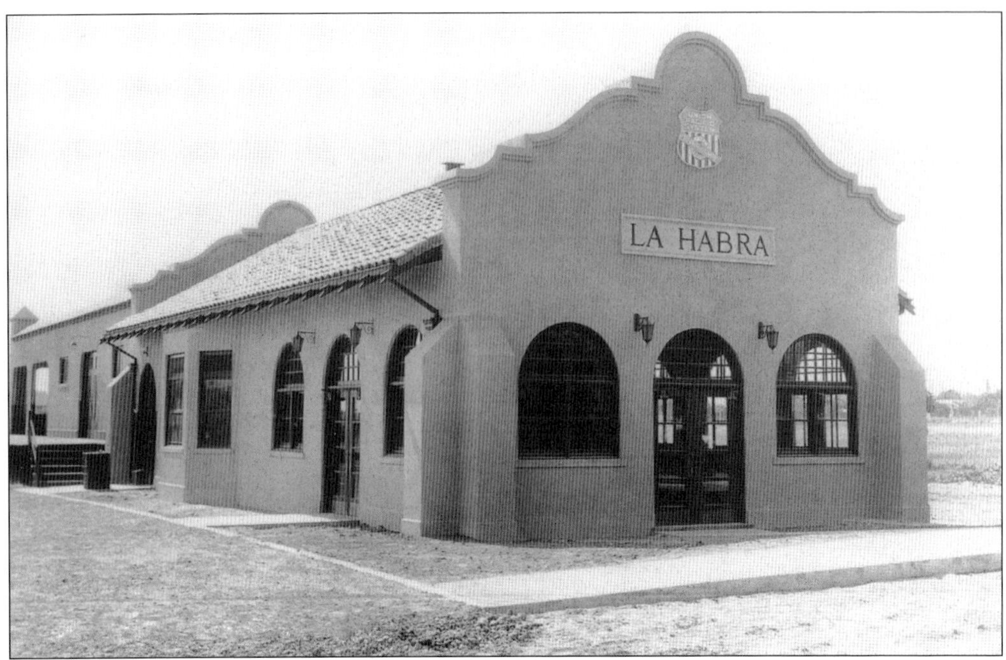

The La Habra Union Pacific Depot was built to provide freight service to the city in 1923 and closed to the public in 1950. The depot sat unused for over 20 years until the historical significance and central location proved to be a perfect fit for housing its current inhabitant, the Children's Museum at La Habra. When the doors of the structure reopened in 1977, it became one of the first children's museums on the West Coast. With the addition of the second wing to the building, the museum offers 10,000 square feet of hands-on exhibits to its young museum-goers and their families. (Both, courtesy of the La Habra Historical Society.)

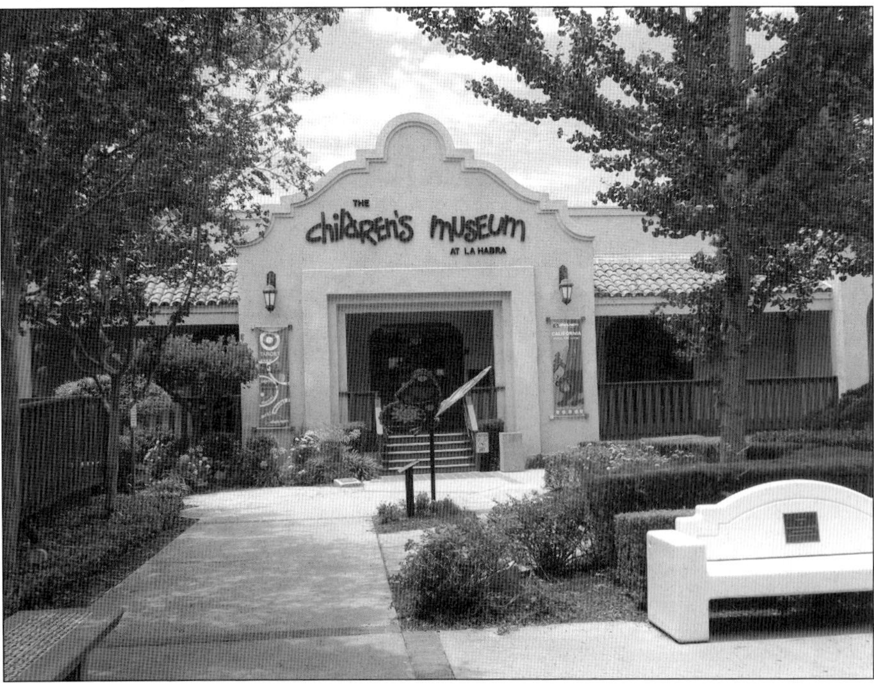

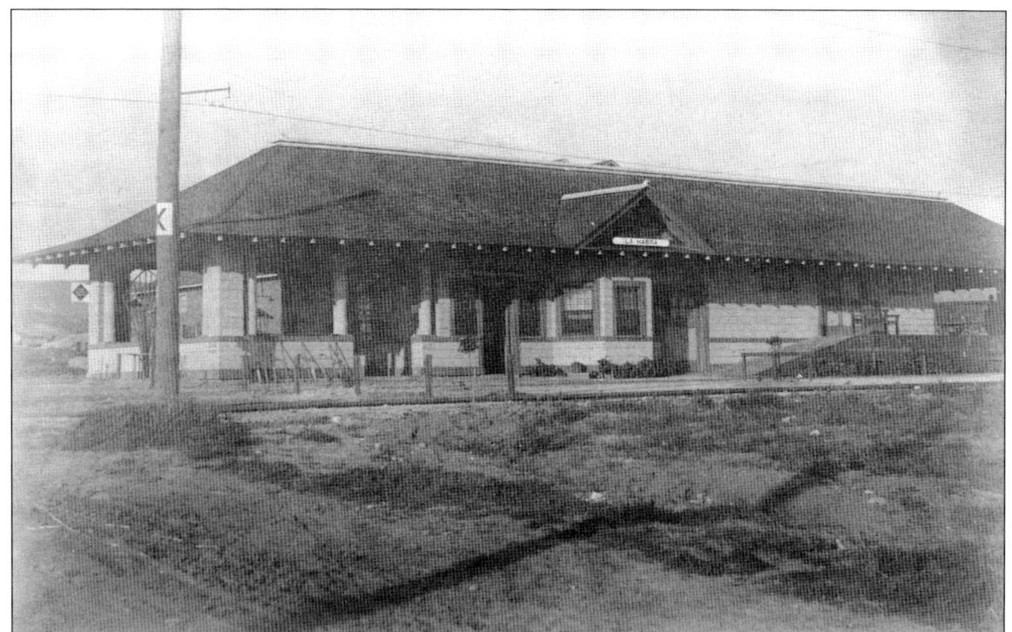

This Pacific Electric depot building symbolized a new era of development in La Habra. Built in 1909, the depot contributed to La Habra's population growth and business boom of the 1920s as crops and supplies could be rapidly transported using the "red cars." The depot was designated a historical landmark in 1976 and was restored to become the La Habra Depot Theatre in 1982. (Courtesy of the Orange County Archives.)

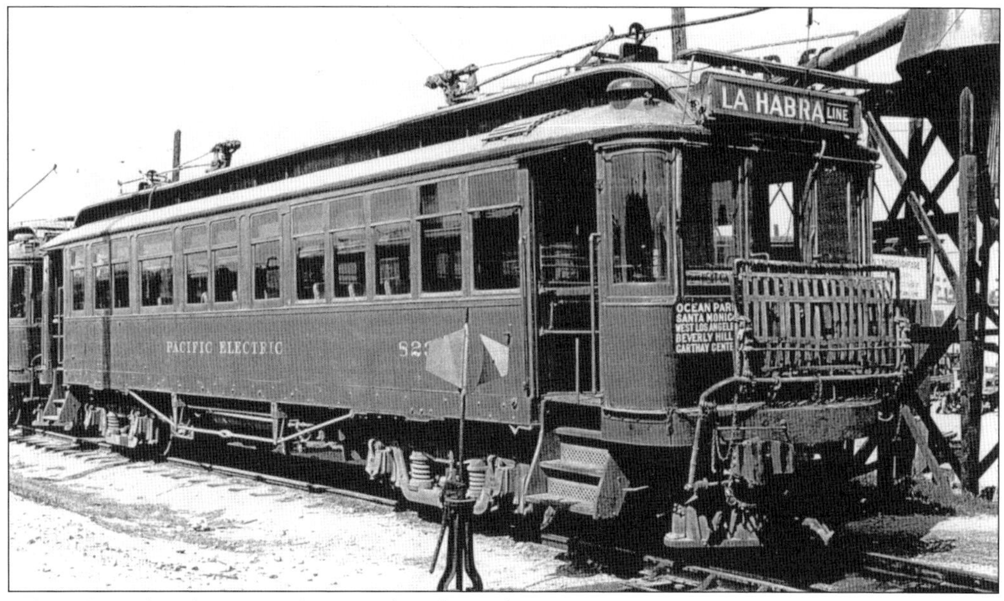

While the railroads in La Habra were largely used for transporting produce, citizens were also able to board to visit nearby cities in a fraction of the time it took before. Passengers could now ride from La Habra to Los Angeles in less than an hour to visit the city or to Huntington Beach in the summer. (Courtesy of the Orange County Public Library.)

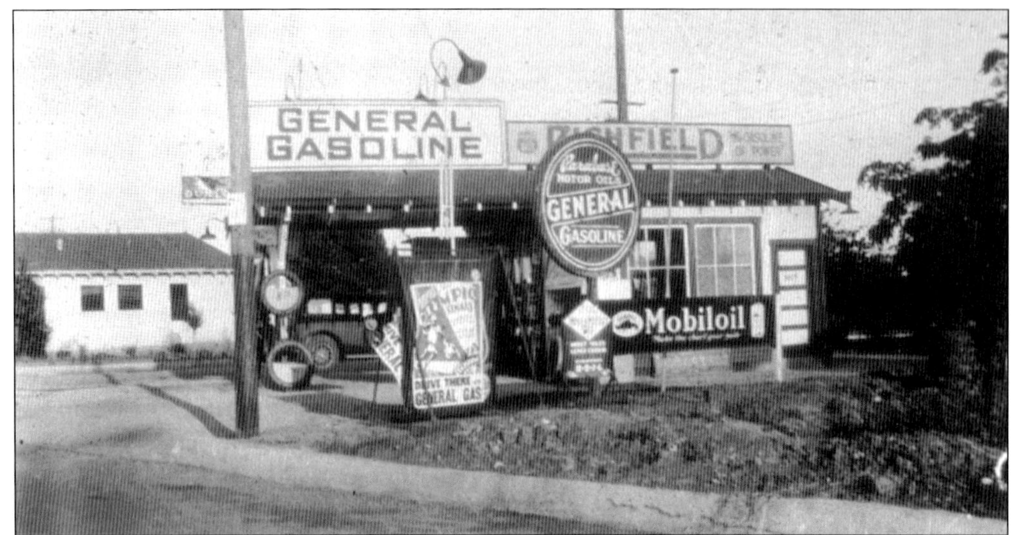

Arriving in La Habra in the 1910s to work in oil fields, Charles Newson became a two-term mayor. He secured Metropolitan Water District membership for the city and oversaw construction of the first city hall in 1935. He also owned the Hart Service Station, later renamed Newson Service Station, shown here in 1928. (Courtesy La Habra Historical Society.)

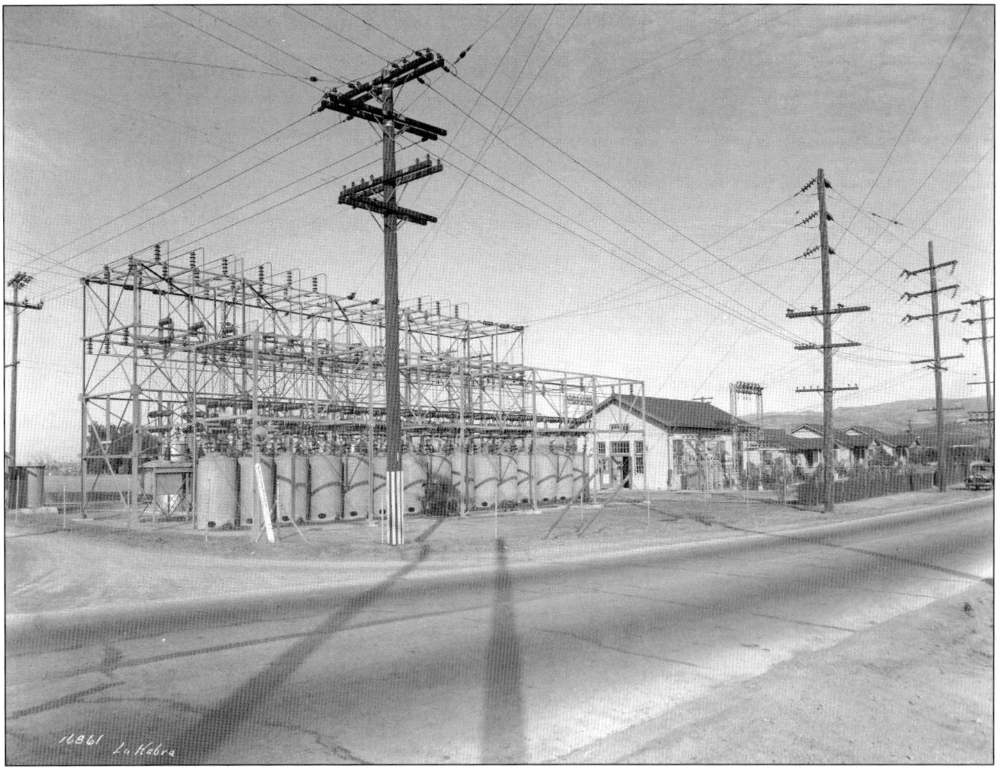

While electric lines passed through La Habra as early as 1902, there is not much evidence supporting their use until a much later date, as consensus on how to finance the expense could not be agreed upon. Substations like this one on the outside of La Habra, photographed on June 7, 1938, controlled and provided service to the city. (Courtesy of Calisphere.)

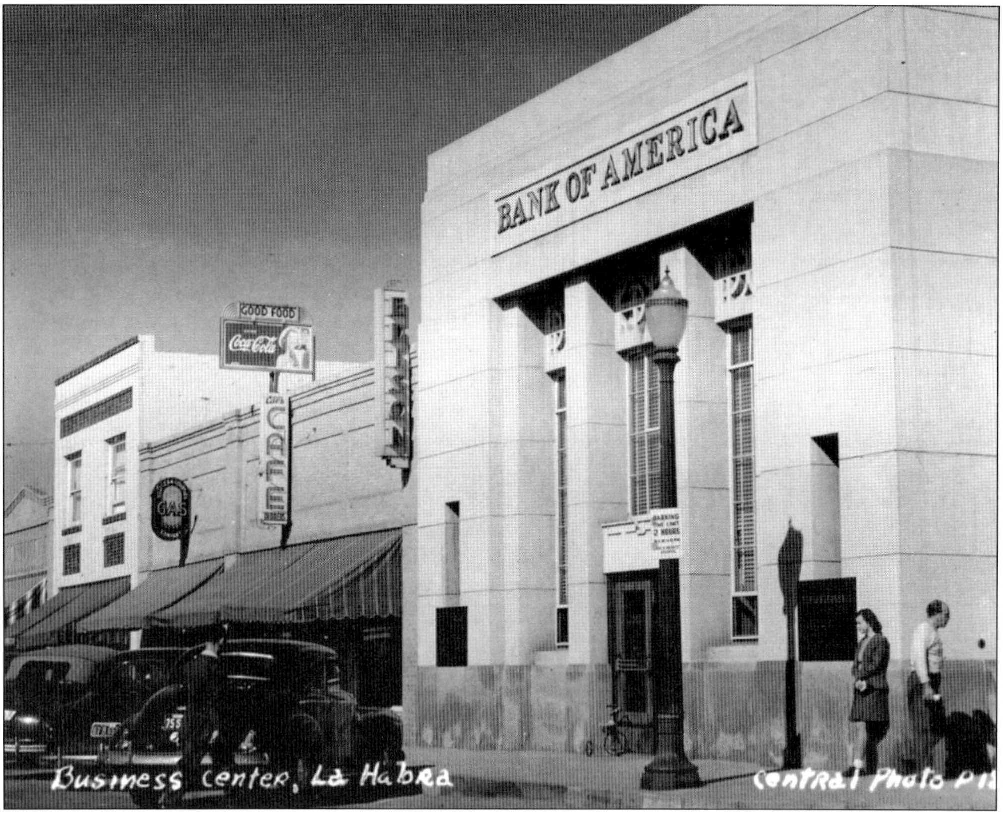

BANK OF AMERICA

Business Center, La Habra

One of the most prestigious buildings in town, the First National Bank opened in 1927 on the northwest corner of what are now La Habra Boulevard and Euclid Street. The picture below documents its opening day, which had much fanfare for the occasion. The manager and bank directors can be seen in the photograph standing to the left of the front doors, which opened onto La Habra Boulevard. The building was purchased by the Bank of America in 1934 and served La Habra for many years before it was demolished in 1982 to widen Euclid Street. (Both, courtesy of the La Habra Historical Society.)

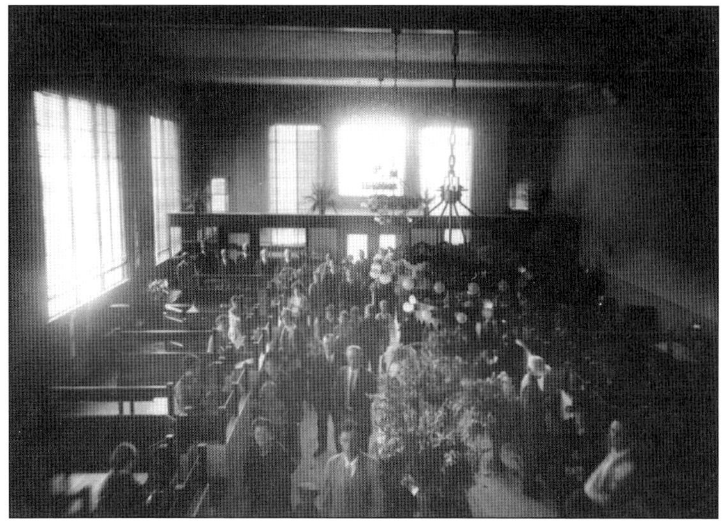

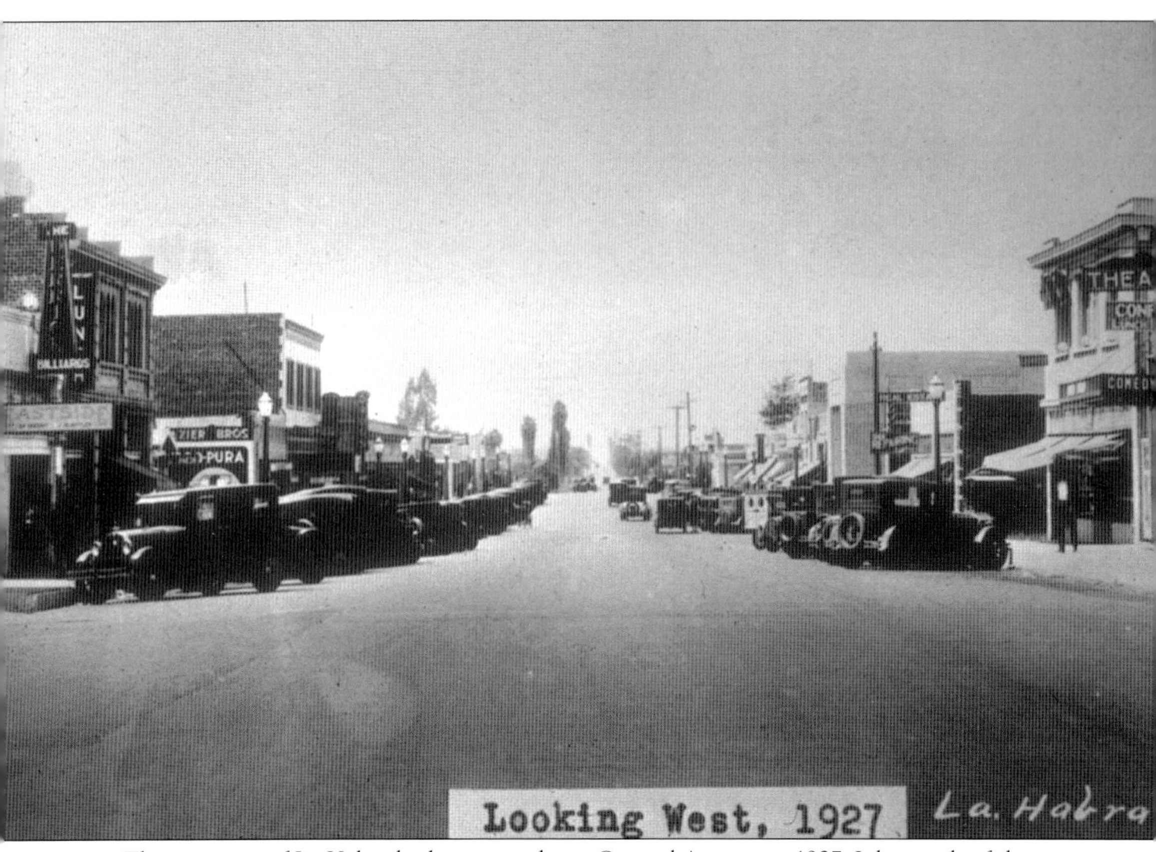

This is a view of La Habra looking west down Central Avenue in 1927. Like much of the country in the 1920s, smaller cities such as La Habra were experiencing a business boom. Many merchants of La Habra at this time raced to bring their storefronts to the rapidly expanding Central Avenue. By 1925, the following establishments were some of many operating in the community: two banks, two drugstores, two hotels, two hardware stores, six meat markets and eight groceries, two clothing stores, four restaurants, two furniture stores, eight garages with service stations, a maternity home, a shoe store, two billiard parlors, a jewelry shop, five real estate firms, a blacksmith shop, a beauty shop, two lumberyards, a bakery, a new dance hall, a fumigation company, an optometrist, and three doctors. This street ran east-west through the center of the newly incorporated city and was later renamed La Habra Boulevard for the stretch that runs through its namesake. (Courtesy of the La Habra Historical Society.)

Three

Revisiting Our Roots in Agriculture

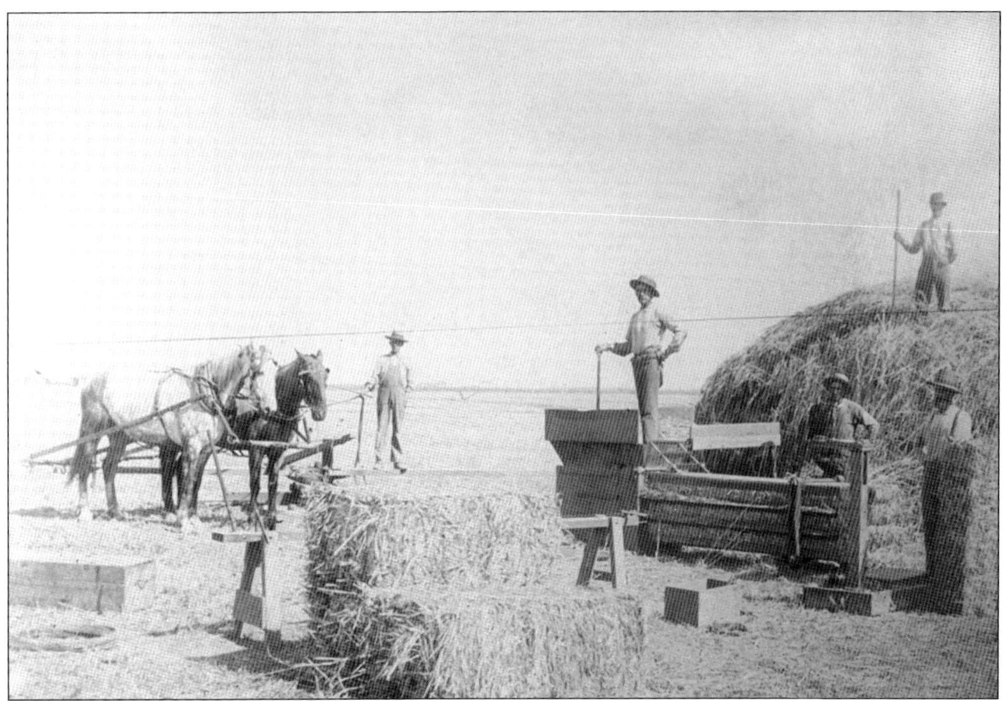

Thomas P. Warne's sons harvest hay on their ranch in 1897. The Warne Ranch once claimed the land around what is now the southeastern corner of Harbor and La Habra Boulevards and held the house he built with his nine sons in 1894, making it the first home built in the village of La Habra. (Courtesy of the La Habra Historical Society.)

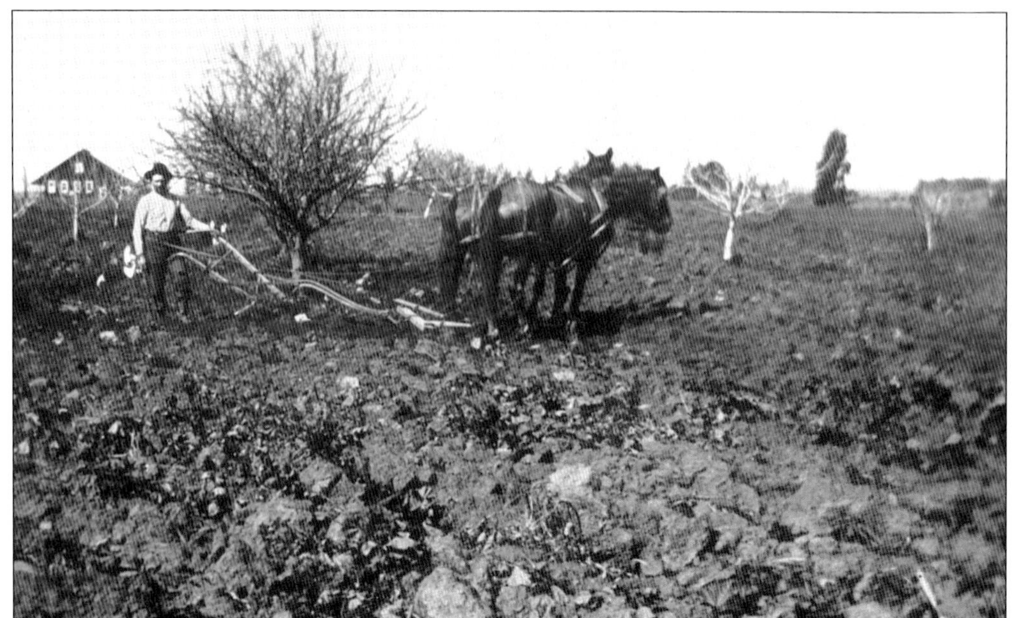

Milton Keeler shows off his horse-drawn plow in 1908. Many early settlers used this method to break up the soil to plant vegetables near their homes. In the background are apricot trees planted by the previous land owner, an English colonizer. Keeler was also one of the first farmers to grow tomatoes in La Habra. (Courtesy of the La Habra Historical Society.)

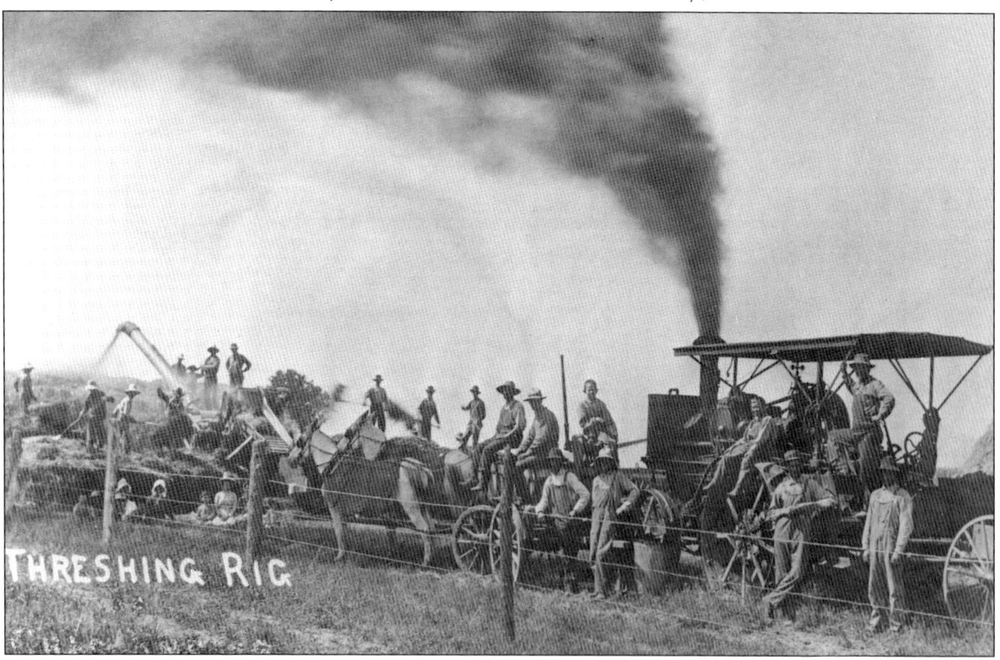

This photograph shows a threshing rig in La Habra around the turn of the 20th century. These large machines that separated grain from straw were critical to the expansion of American agriculture. They were typically too expensive for individual farmers, so neighbors would often go in together to buy and share the equipment, a practice very in line with the city's motto of "a caring community." (Courtesy of the La Habra Historical Society.)

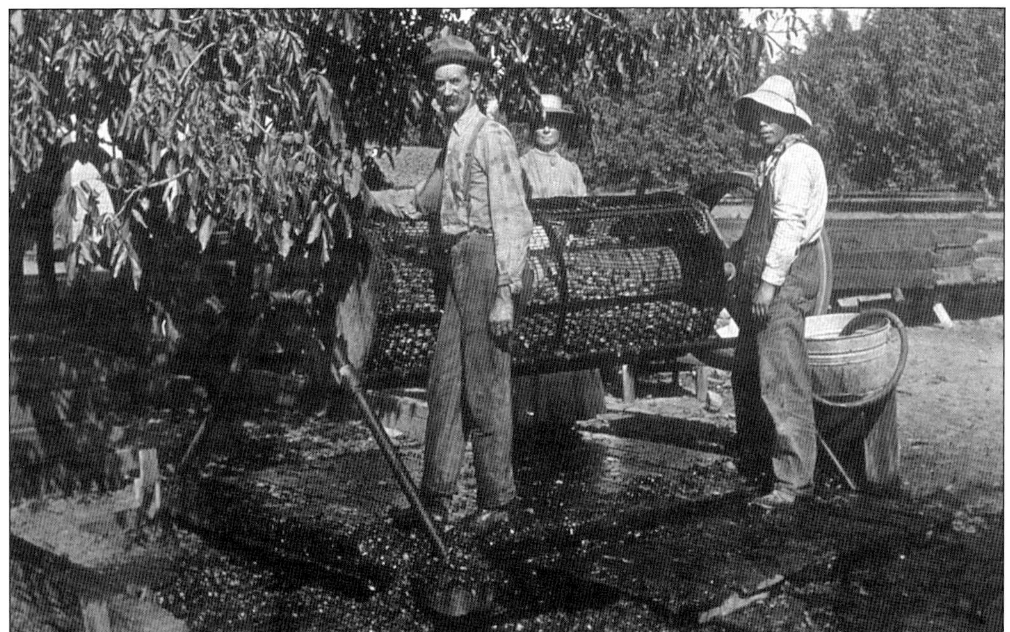

Farmhands operate a walnut hulling machine at Smith Ranch. S.M. Smith began growing walnuts in 1895 and had over 100 acres dedicated to the trees. The hulling machine was a drum-like machine that was turned to remove the leathery green outer layer of the walnut fruit. (Courtesy of the La Habra Historical Society.)

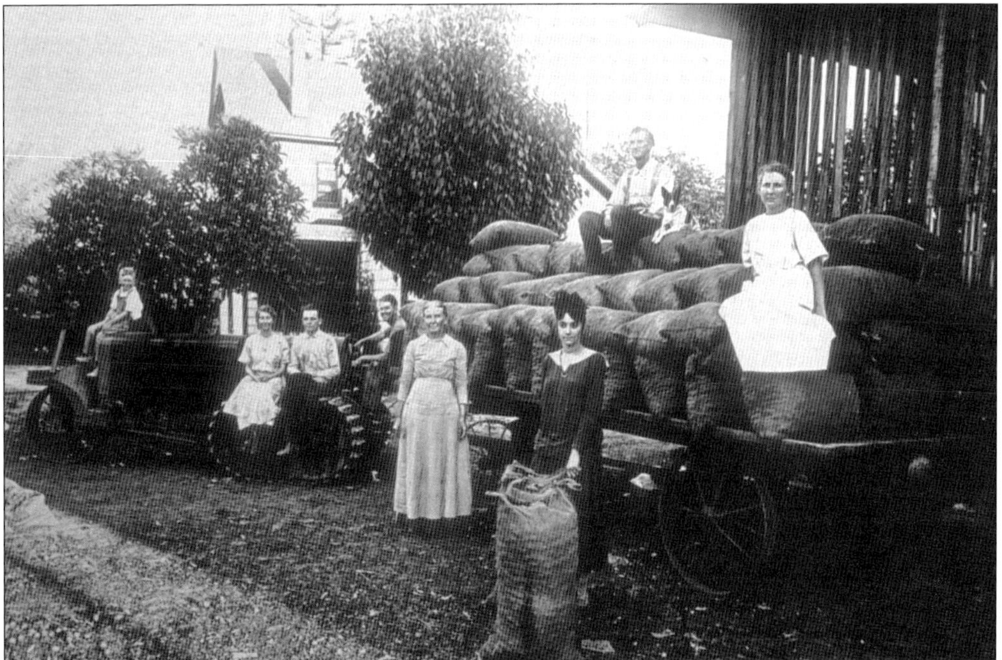

After the walnuts dried, they were packed in burlap bags and transported to the packinghouse for bleaching, grading to determine their price, and bagging. In this photograph from 1910, S.M. Smith sits atop his harvest with his dog, and his wife stands in a long dress beside the flat bed. (Courtesy of the La Habra Historical Society.)

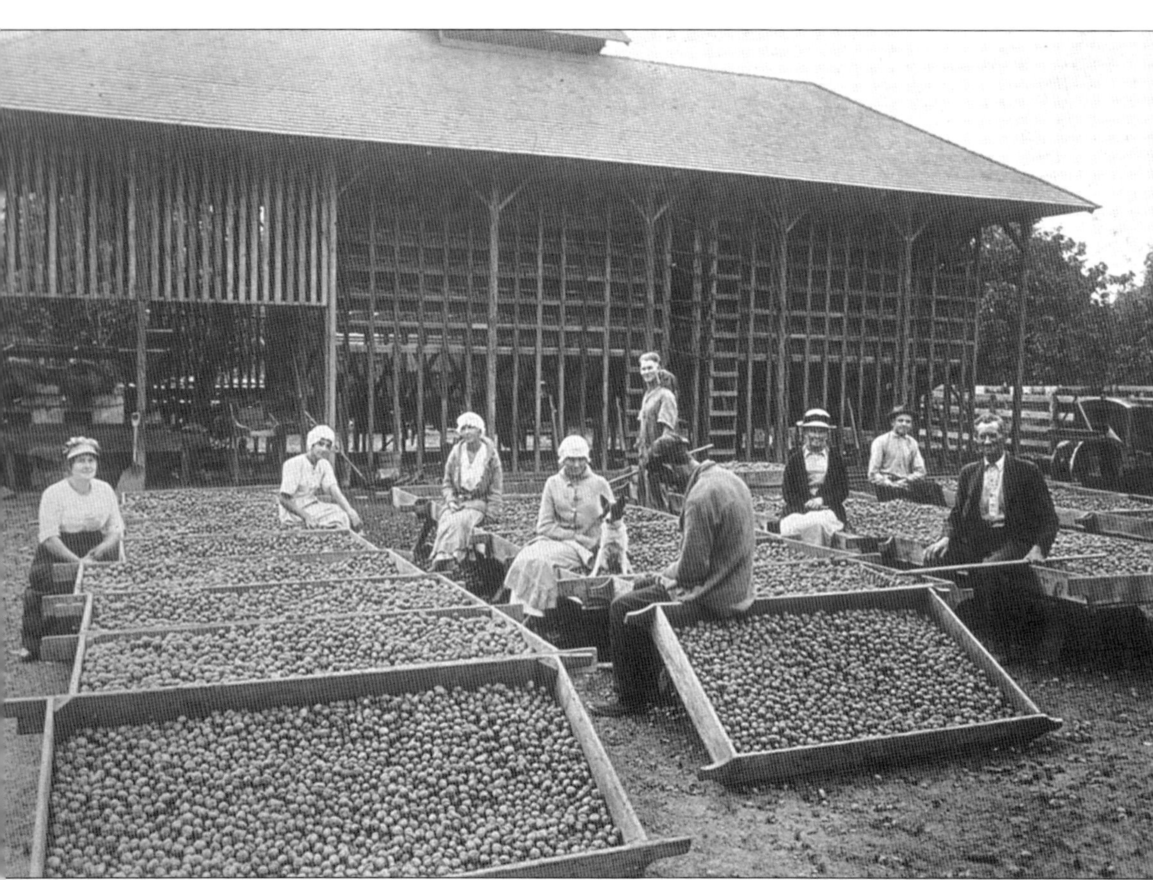

Walnuts were left to dry in the sun in large trays before they were shipped to packinghouses, as seen in these pictures from Smith Ranch. At the time, the responsibility for cleaning and drying the walnuts fell on the ranchers instead of the packinghouse. The walnut industry boomed in La Habra from 1916 to 1924, with its peak year recognized as 1919, when a total of 300 tons of walnuts were shipped from the La Habra Walnut Growers Association. Many Mexican families migrated to the region during this era to work in the walnut groves, and each year, Smith Ranch employed some of these temporary laborers to help with their harvest. (Courtesy of the La Habra Historical Society.)

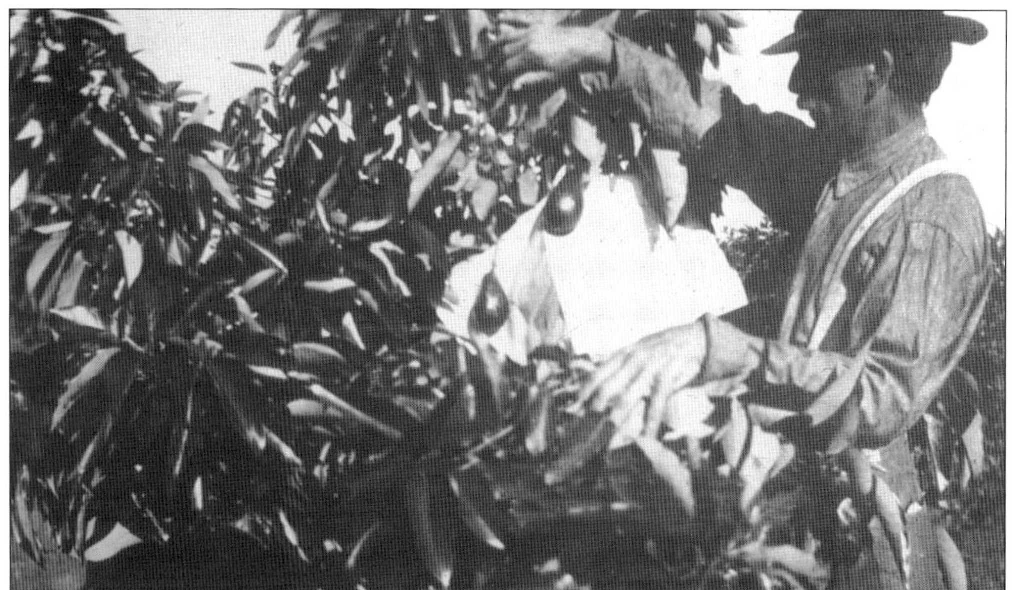

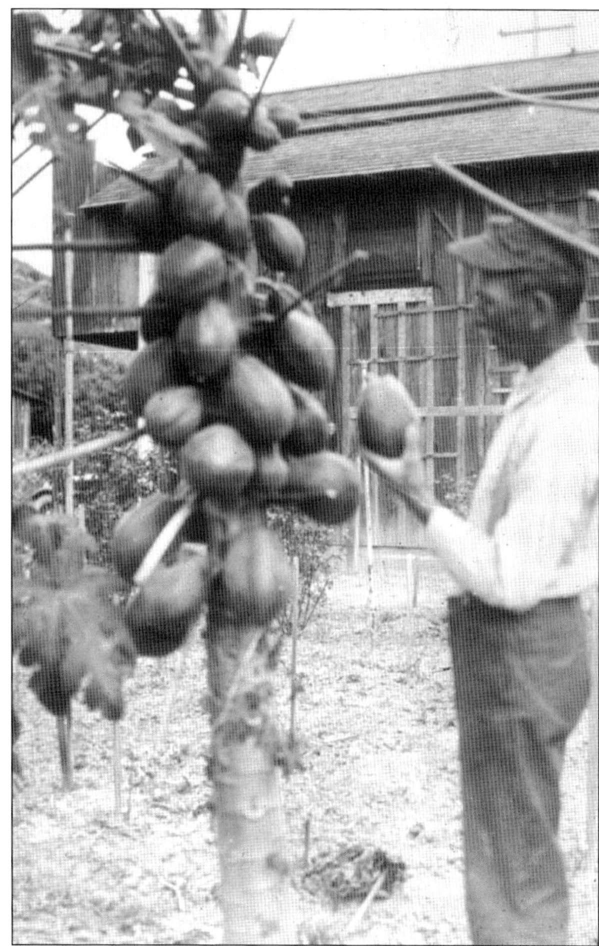

The first avocados to be grown commercially in Orange County came from the trees of George W. Beck's ranch; he is pictured above with one of those trees in 1910. This ranch was located off of Cypress Street, north of Whittier Boulevard, and was nicknamed "the Garden of Fruits." In addition to avocados, Beck grew many types of subtropical fruits not native to the region, such as bananas, kumquats, guavas, and papayas, as pictured in the photograph to the right. Beck planted most of these fruits out of personal interest rather than for profit. (Both, courtesy of the La Habra Historical Society.)

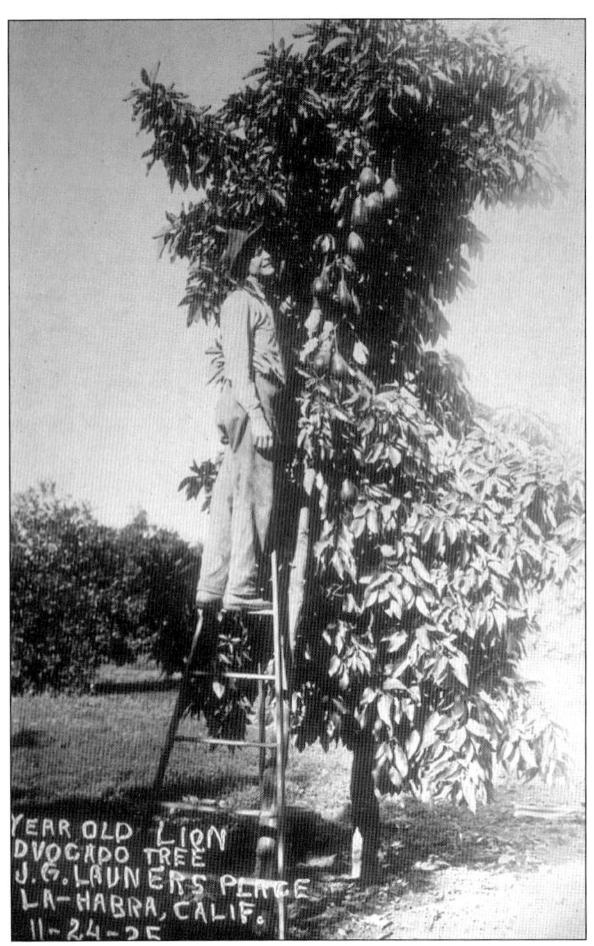

Avocados are a long-standing agricultural icon of La Habra, as shown by the city's first mayor, John Launer, harvesting some from his Lyon variety avocado tree on November 24, 1925. A wider shot of the Launer family home and surrounding farm at Central Avenue and Main Street can be seen below 20 years earlier in 1905. The Hass avocado originated from a seed that was planted by Rudolph Hass in his La Habra Heights backyard, and while this historic tree died in 2002, the plant is still synonymous with the agricultural history of La Habra. (Both, courtesy of Orange County Archives.)

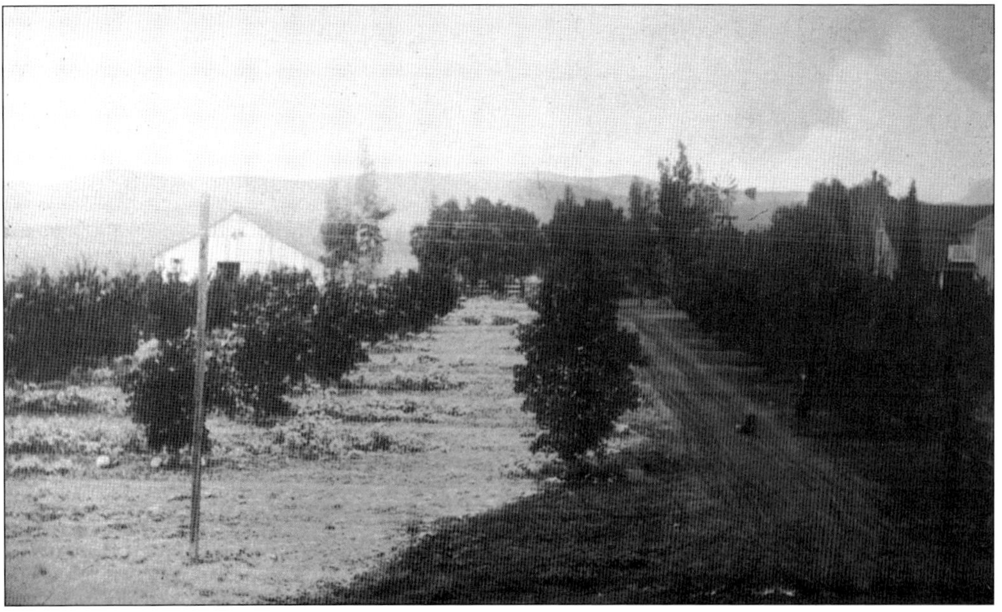

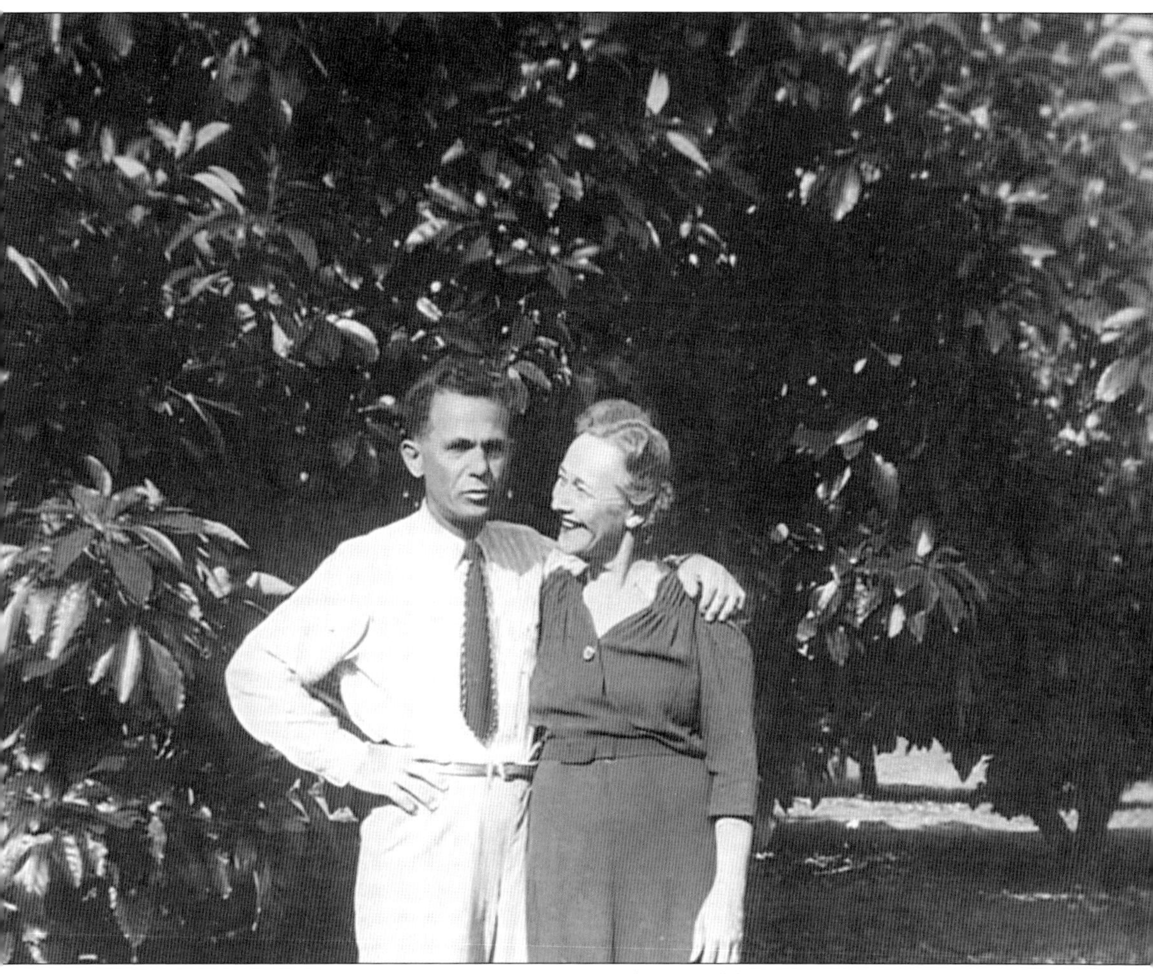

Rudolph and Elizabeth Hass pose in front of a farm of their world-famous Hass avocado trees. After seeing a magazine ad depicting money growing on an avocado tree, Hass began grafting plants out of different seeds purchased from Whittier nurseries. Through these efforts, he created the smooth and tasty Hass avocado. He recognized the variety's potential once his family and neighbors began complimenting the rich taste and longer shelf life and patented the Hass avocado in 1935. The fruit today accounts for the majority of avocados grown and consumed worldwide, surpassing the Fuerte variety. The Hass mother tree, planted in La Habra Heights in 1926, died of root rot after 76 years in 2002. (Courtesy of the California Avocado Commission.)

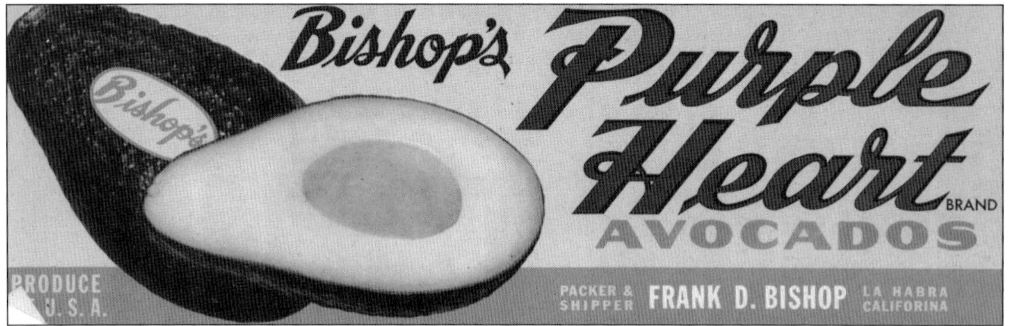

Bishop's Purple Heart Avocados crates contained labels such as the one seen here. Bishop's was a prominent local shipper of avocados in La Habra for many years and, like many shippers at that time, took out ads in the *La Habra Star* newspaper offering high prices directly to growers for their avocados instead of selling through the Avocado Growers Association. (Courtesy of the La Habra Historical Society.)

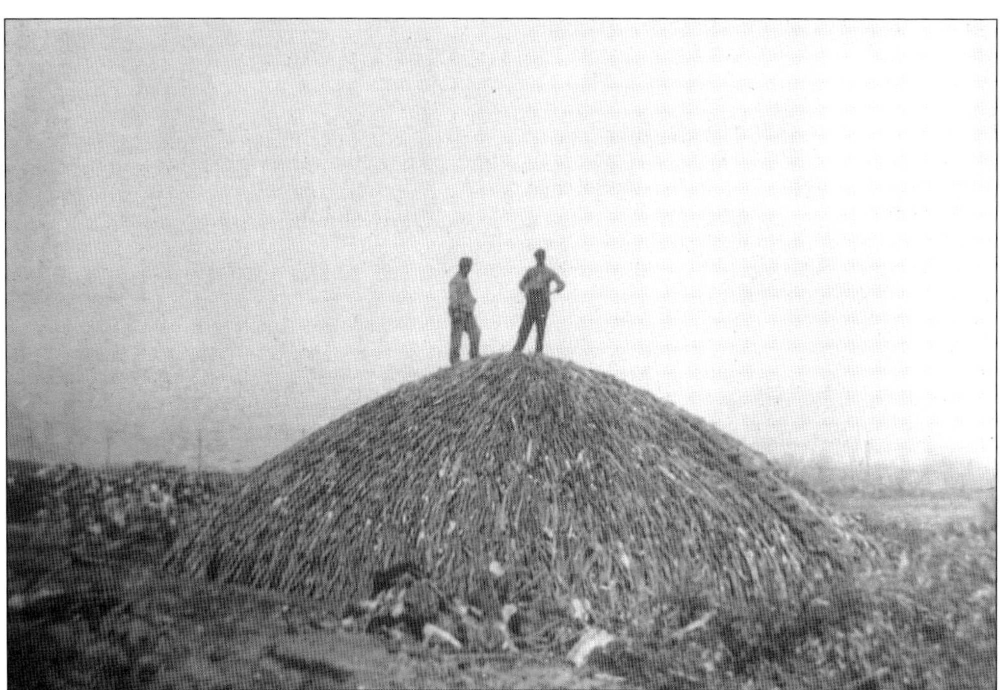
In 1917, two men stand on a pile of olive wood that was stacked up to burn into charcoal. Olive wood is known to be slow-growing and is often planted in areas with little water, making it an ideal fit for the mostly dry climate of La Habra when irrigation was still rather new to the area. (Courtesy of the La Habra Historical Society.)

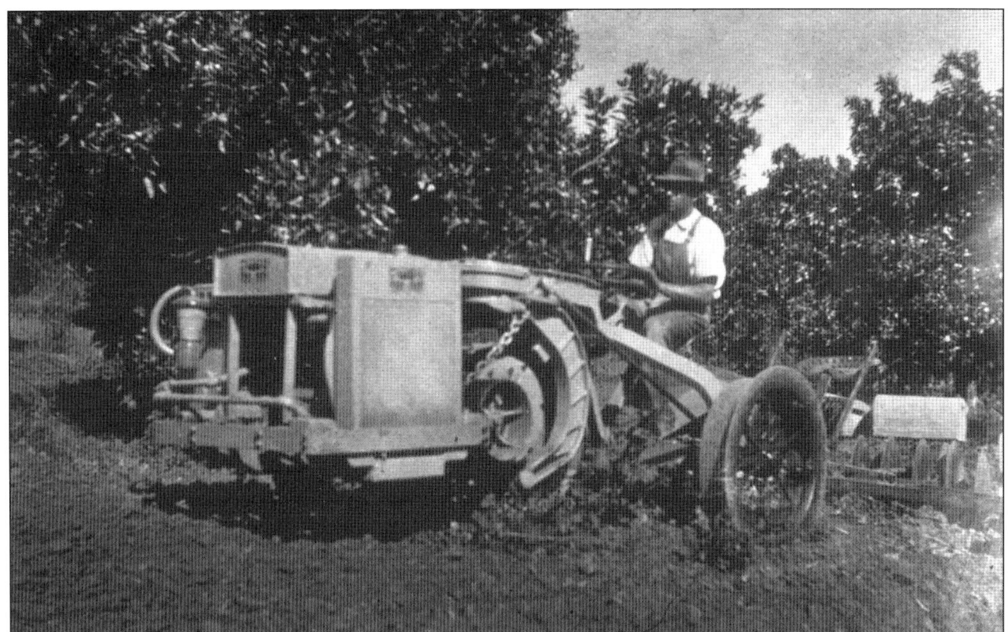

Edgar Leutwiler, an early settler to the area, sits atop a tractor pulling a disc through a young citrus grove on the Leutwiler Ranch around 1915. This equipment was used for levelling heavy soil, uprooting weeds, and preparing the ground for planting seeds. (Courtesy of the La Habra Historical Society.)

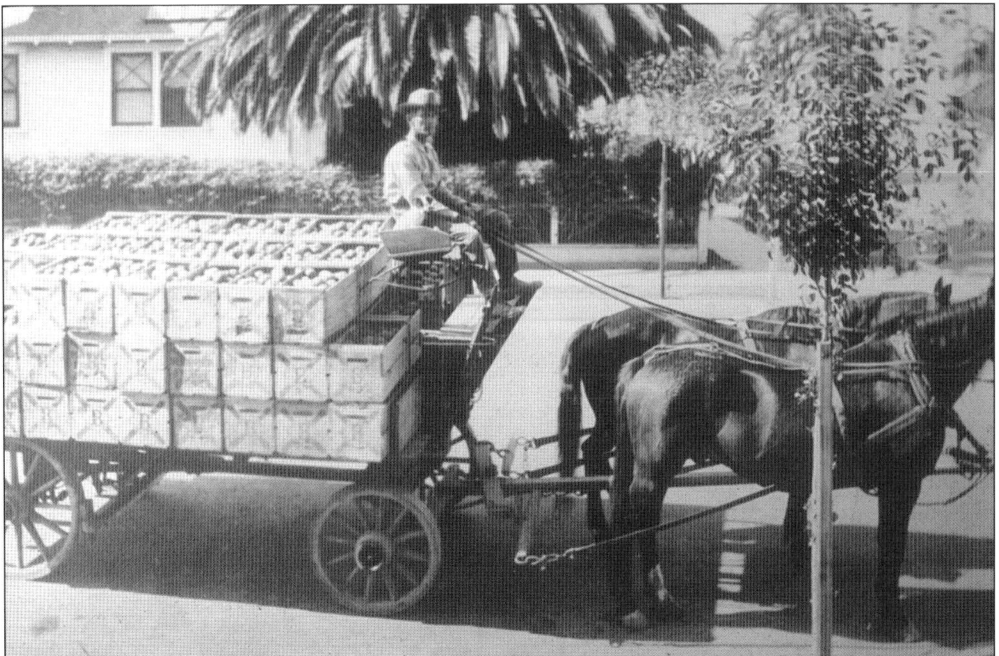

Riley "Bud" Warne hauls a load of citrus, worth $220 then and upwards of $5,000 now, from La Habra to a packinghouse in Fullerton. Warne was a well-known local citrus hauler and, with his horses and wagon, transported thousands of boxes of La Habra citrus, for which he was paid by the box. (Courtesy of the La Habra Historical Society.)

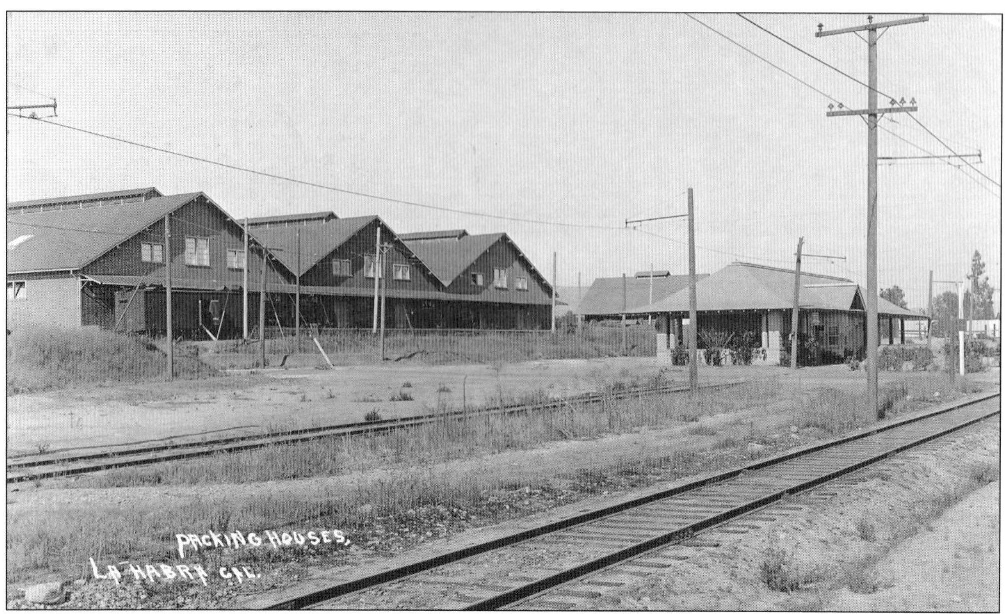

The La Habra Citrus Association was formed in 1911 and became part of the California Fruit Grower's Exchange. Seen here are the inside and outside of the La Habra Packinghouse, which was the first packinghouse in the city dedicated to citrus fruits. The citrus fruit was graded at the packinghouses, and local growers who were very proud of their produce would name the premium grades of both lemon and oranges "La Habra" and the second grades "Reliable." Both of these grades were "Sunkist," the generic term for the best-quality fruit packed by the California Fruit Growers' Exchange. (Both, courtesy of the La Habra Historical Society.)

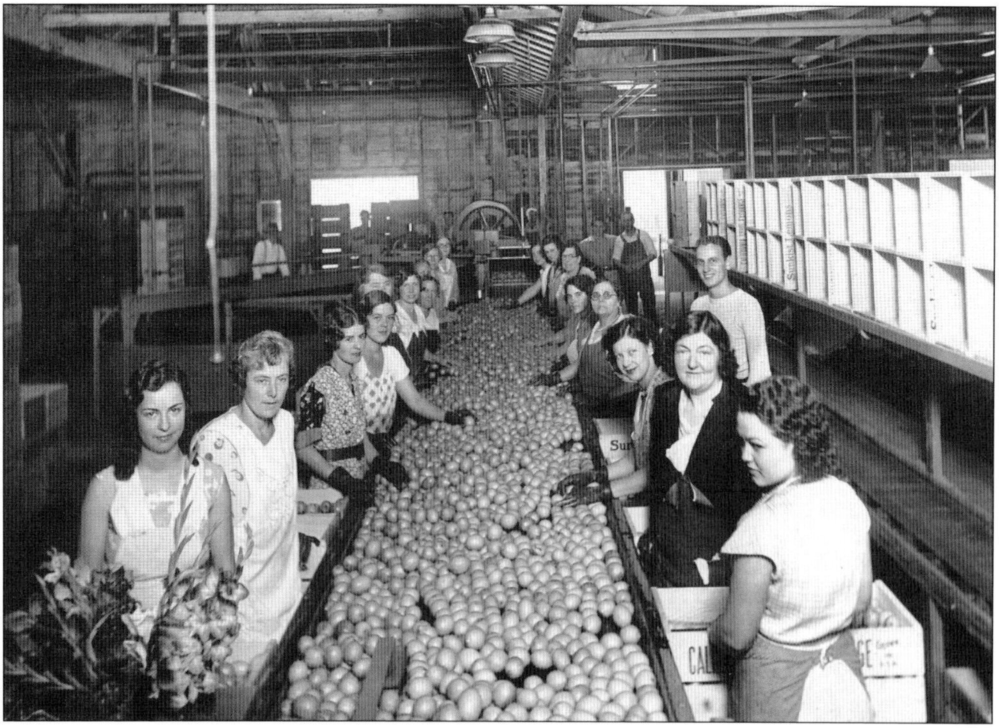

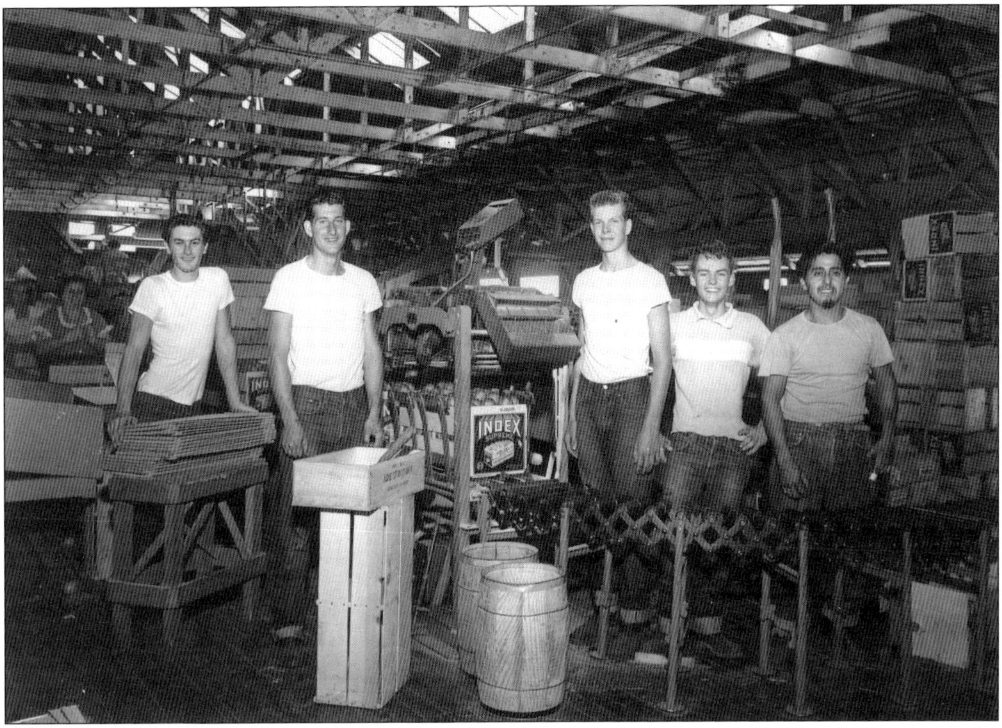

Of all the early citrus fruits grown in La Habra, lemons were the most successful. The prototype for early lemon ranches in this area was the C.W. Leffingwell ranch, Lemonita, located near the present Whitwood Shopping Center, and it was a completely self-contained growing-to-shipping operation. Above, a group of young men works at the Index Superior Lemon Packing House. Below, a young Edward "Guero" Velasquez poses for a photo op on his first day at work in the orange groves around 1950. During this time, every able-bodied individual in the community was recruited to work in the industry in some way, yet the local labor was not enough. Mexican workers, largely from Jalisco, Chihuahua, and Sinaloa, were recruited by the California Fruit Growers' Exchange. (Above, courtesy of the La Habra Historical Society; below, courtesy of Edward Velasquez.)

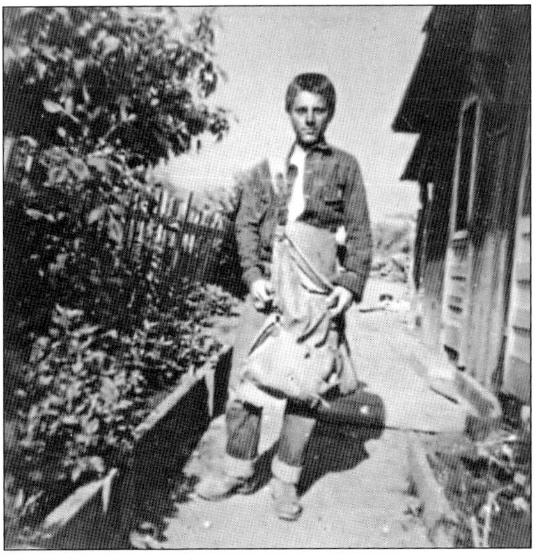

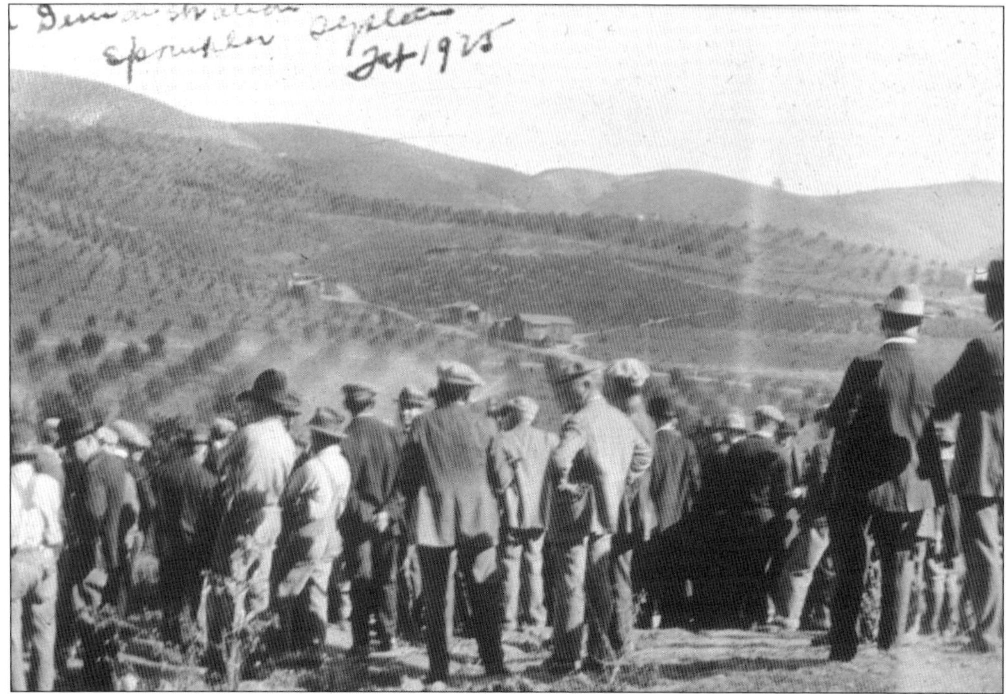

A large crowd watches a sprinkler demonstration for a farming application at Tresslar Ranch on Fullerton Road in the city of Fullerton. Many La Habran farmers were among the spectators, hoping to make their own agricultural operations more efficient or perhaps just for a bit of fun in seeing the latest in new technology in Southern California. (Courtesy of the La Habra Historical Society.)

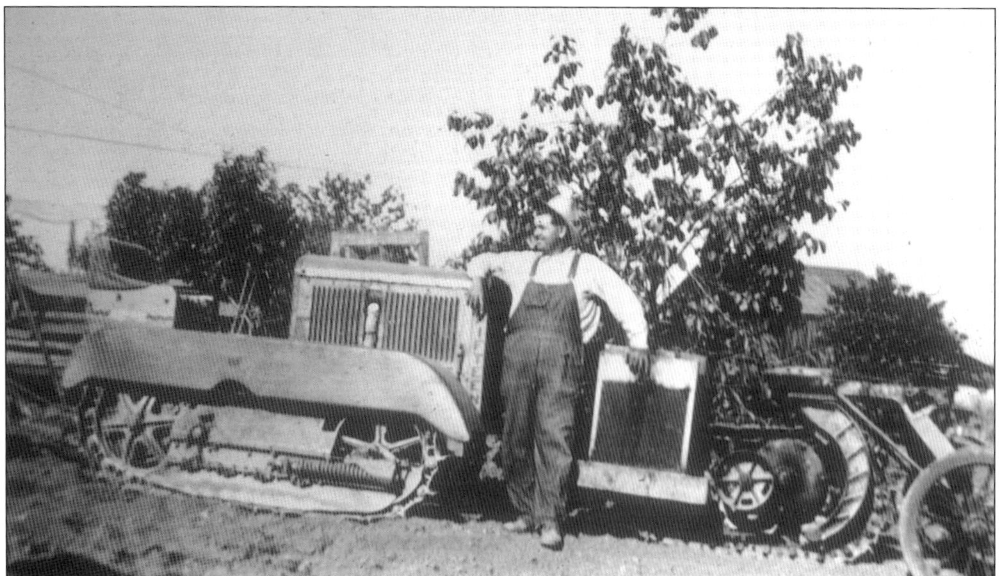

Farmer Harrison Reynolds proudly poses in front of a tractor, which was an impressively large model for the time of this photograph in 1935. Reynolds was also a member of the La Habra Old Settlers Historical Society, serving as the group's vice president in 1967. (Courtesy of the La Habra Historical Society.)

Four

LIFE IN THE MIGRANT CAMPS

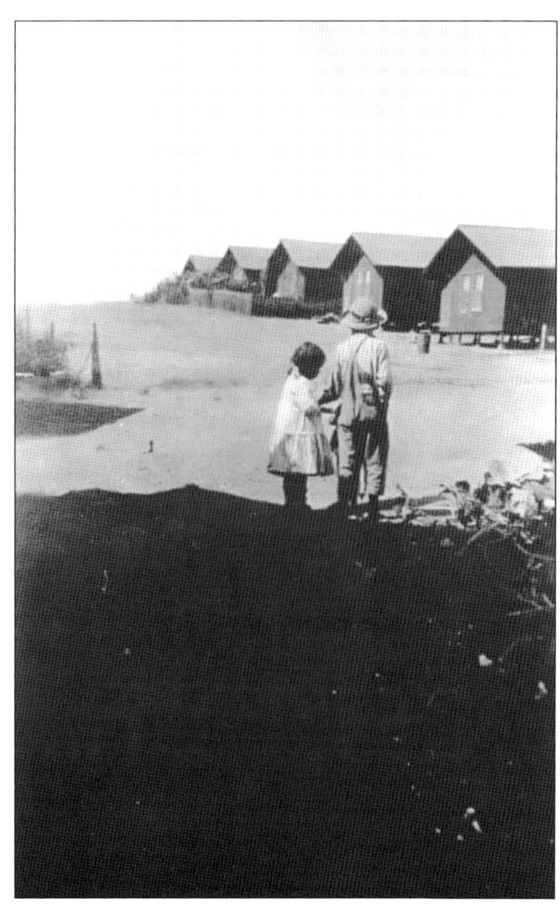

The first migrant camp in La Habra was built by the La Habra Citrus Association in 1916. This initial development became known as Campo Colorado; here, some Mexican immigrant workers stayed in groups of small buildings that became known as "Little Mexico." Most of the workers lived alone or squeezed into the one-room structures with their families. (Courtesy of the La Habra Historical Society.)

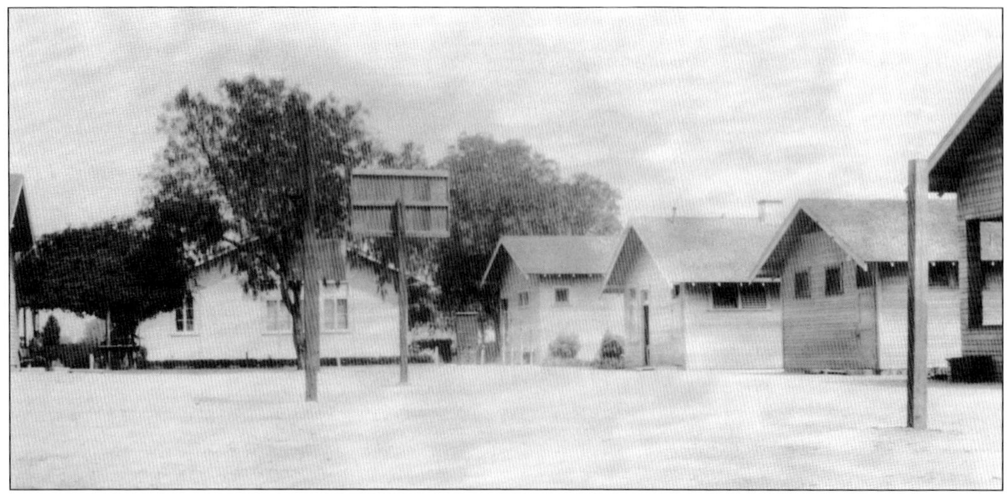

The West Side School was built in 1920 as a temporary school to educate the Spanish-speaking children of the citrus workers brought to La Habra from Mexico by the La Habra Citrus Association. It was intended to serve as both a school and a community center for the Mexican immigrant families to help them adjust to life in the United States. (Courtesy of the La Habra Historical Society.)

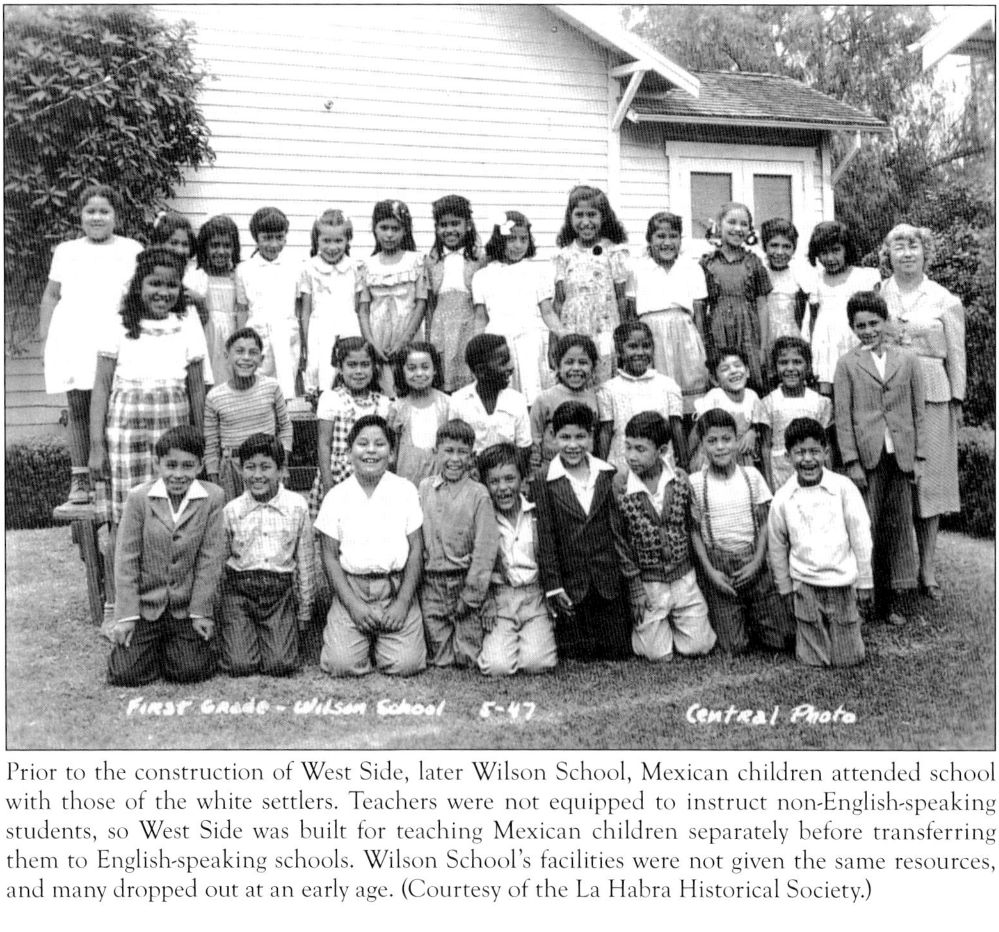

Prior to the construction of West Side, later Wilson School, Mexican children attended school with those of the white settlers. Teachers were not equipped to instruct non-English-speaking students, so West Side was built for teaching Mexican children separately before transferring them to English-speaking schools. Wilson School's facilities were not given the same resources, and many dropped out at an early age. (Courtesy of the La Habra Historical Society.)

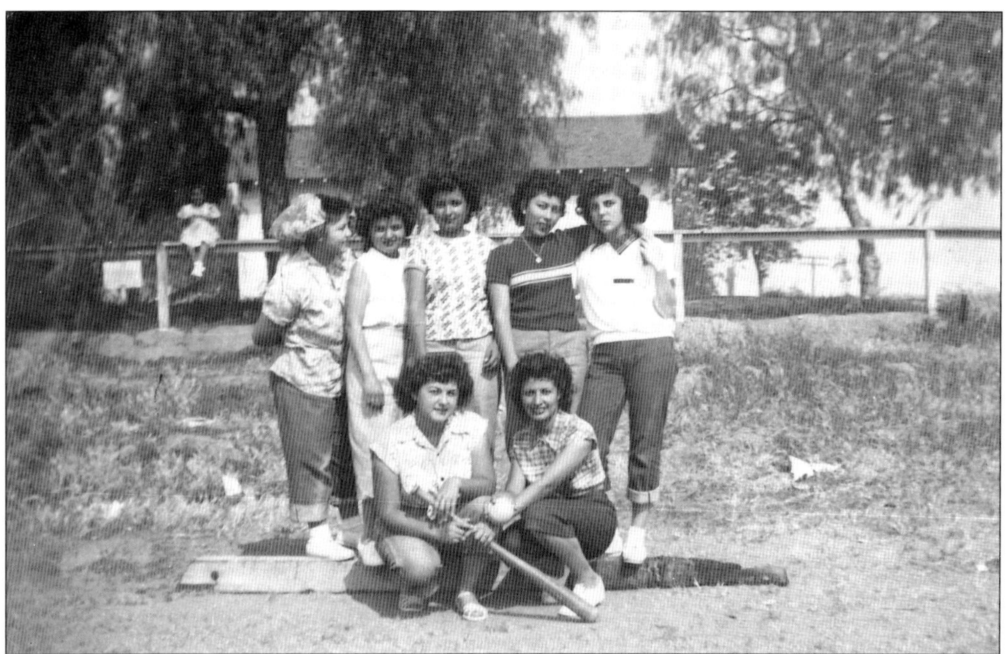

A girl's baseball team from Wilson School poses for a team photograph in the dirt lot where they played in 1950. The school was built at Second Avenue and California Street, near present-day Portola Park, where many baseball games, such as the ones these girls played in, are still played today. (Courtesy of Rachel Torres.)

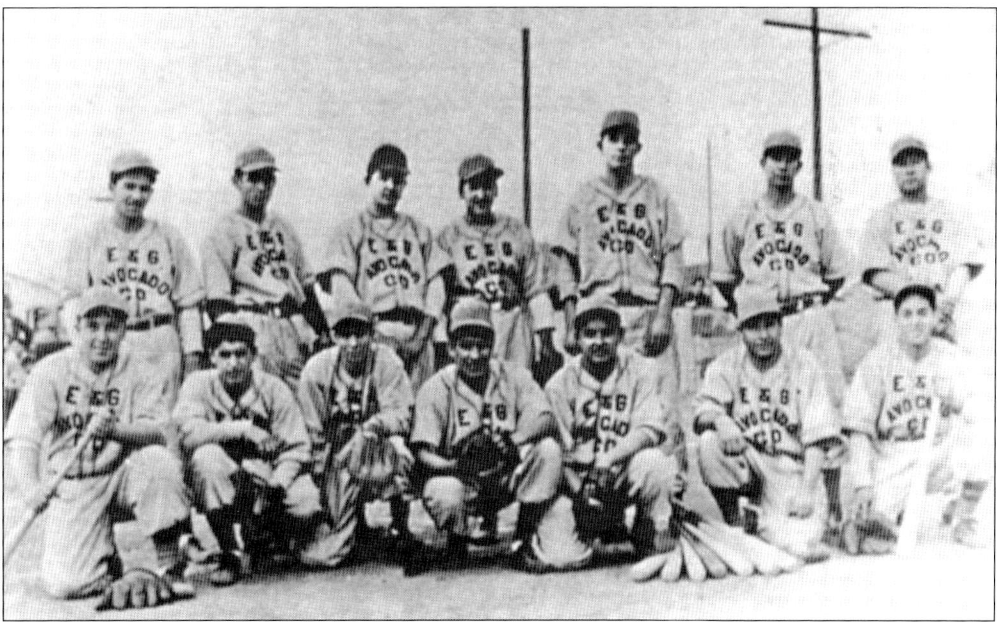

During the height of the agricultural industries in Orange County, most migrant camps sponsored at least one baseball team. These teams played in what became known as the citrus league, and while players and fans were invested in the game, growers thought of these teams as an investment in creating better, more devoted workers for the packinghouse they played for. (Courtesy of Tony M. Gonzalez.)

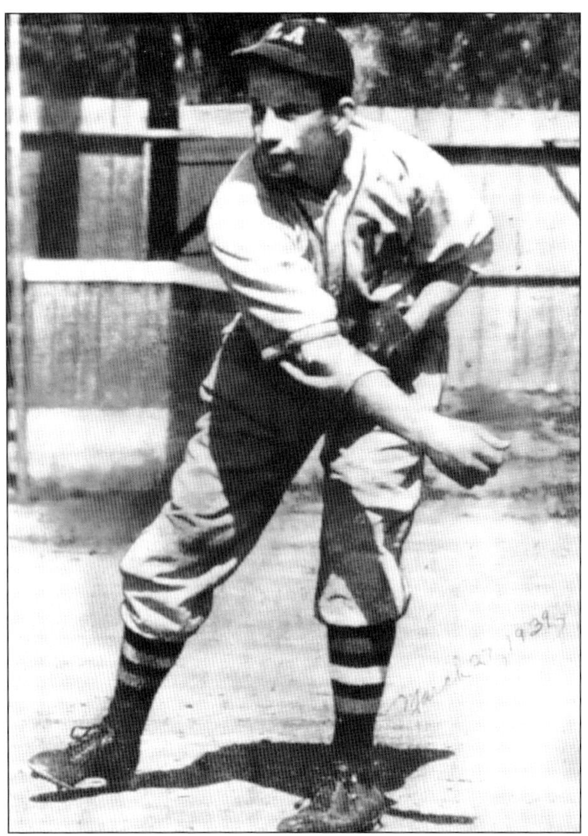

One of the most famous players to emerge from these leagues was Jesse Flores. Flores played third base and pitched for La Habra's Los Juveniles before signing with the Chicago Cubs in 1938, making him the third player of Mexican descent to play in the big leagues. He later played for the Philadelphia Athletics and the Cleveland Indians and became a respected Minnesota Twins scout. (Courtesy of Rose Espinoza.)

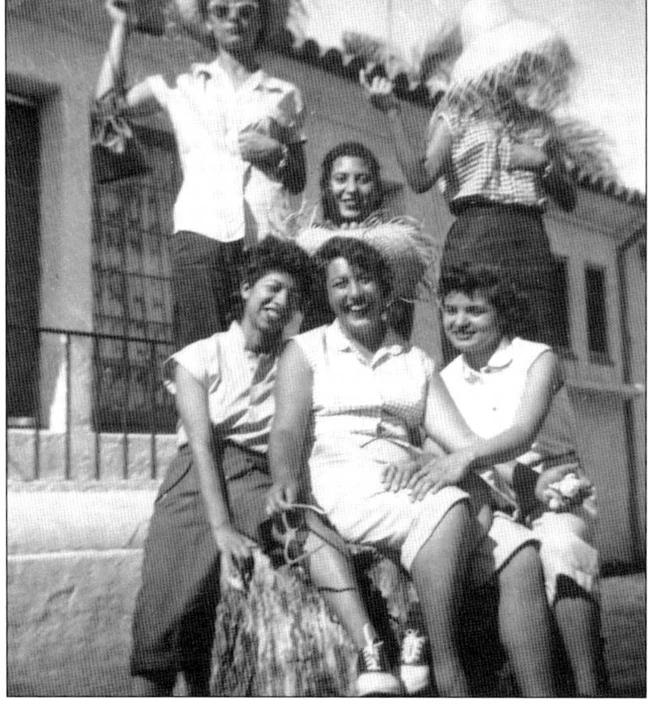

High schoolers at La Habra's Wilson School celebrate during the first Corn Festival in 1949, not knowing the celebration would become synonymous with La Habra. The Corn Festival appealed not only to the city's midwestern transplants but also to the Mexican American immigrants who shared an ancestral home with the yellow crop. (Courtesy of Rachel Torres.)

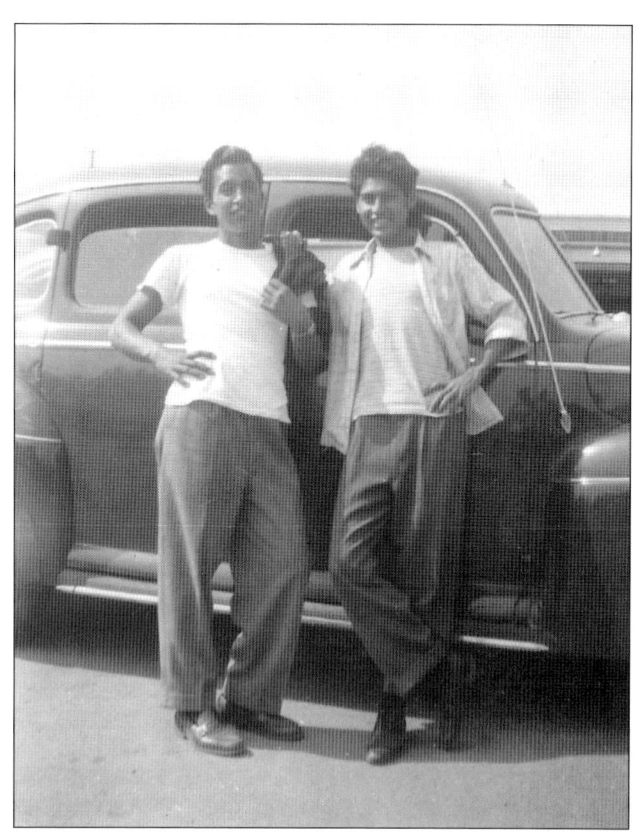

Johnny Yosas (right) and an unidentified friend lean against a car in the late 1940s. Below, his son Johnny Yosas Jr. poses with a view of Valencia Street in the background in 1964. During the 1940s, many Mexican Americans in Southern California found other employment in the wartime economy. At this point, the citrus industry also began the bracero program of using Mexican contract laborers instead of the existing immigrant workforce, which had begun efforts to unionize and bargain for higher wages. When the groves began to disappear completely with suburbanization, many of the families that had lived in the camps remained in La Habra. (Both, courtesy of Rebecca Duarte.)

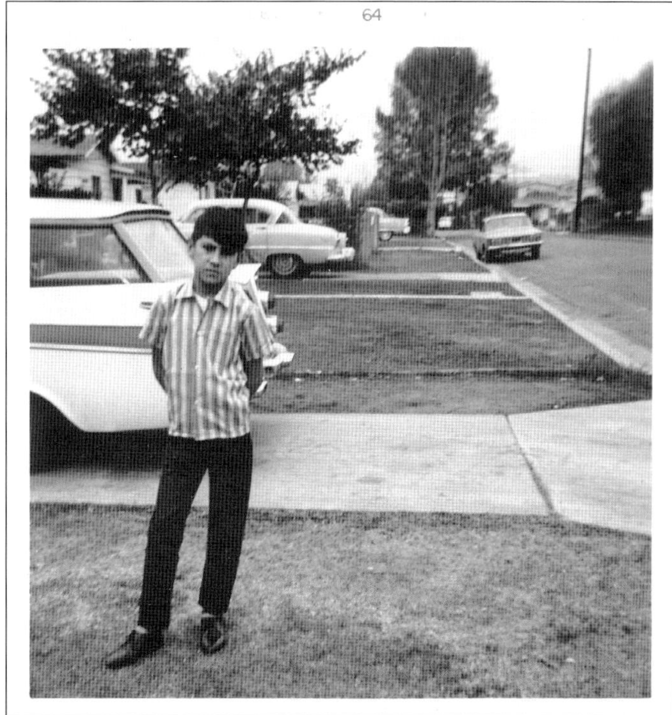

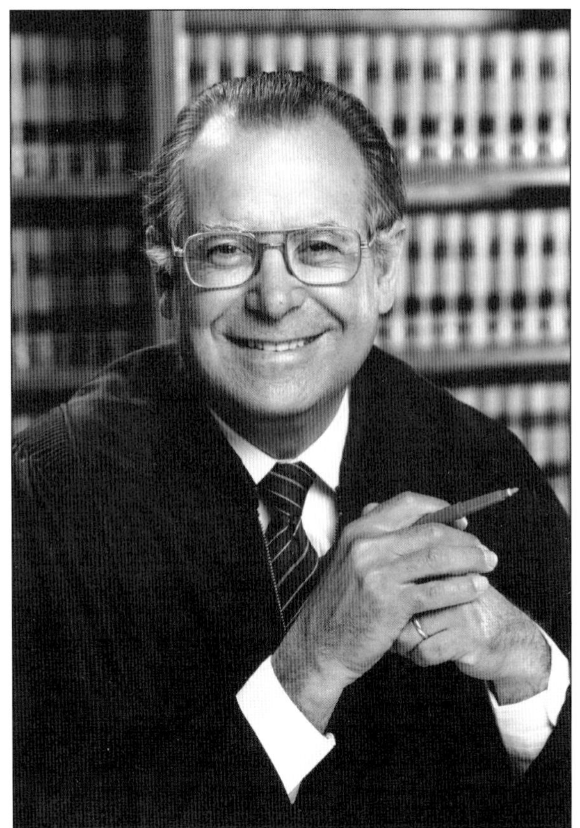

One of the most famous graduates of the now-shuttered Wilson School was Cruz Reynoso, the first Latino state supreme court justice in California. Reynoso moved to Alta Vista in 1938 and began working in the orange groves in addition to attending school when he was eight years old. He was also a veteran of the US Army, appointed to the Congressional Select Commission on Immigrant and Refugee Policy by Pres. Jimmy Carter, and taught at UCLA's School of Law, amongst many other accomplishments. In 2000, Reynoso was awarded the Presidential Medal of Freedom, the highest civilian honor in the United States, by Pres. Bill Clinton. (Left, courtesy of Rose Espinoza; below, courtesy of University of California, Davis.)

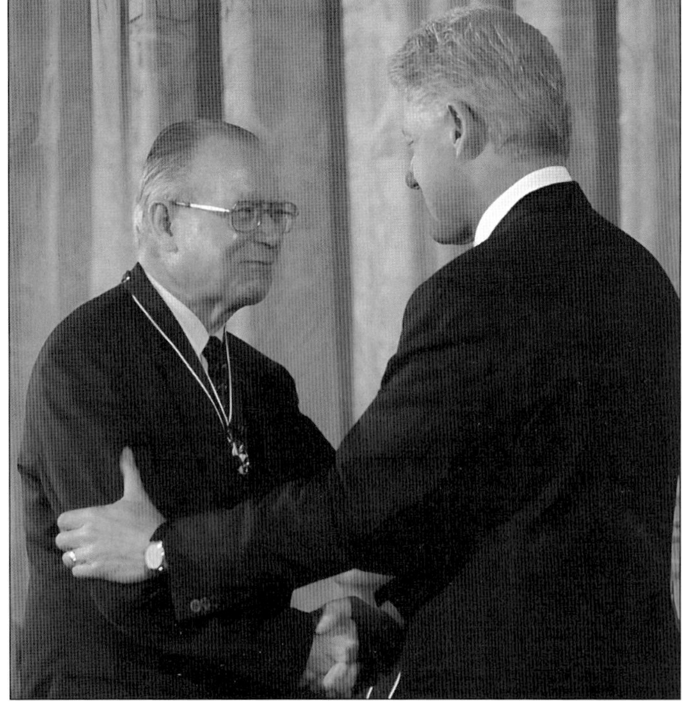

Five

Economic Evolution

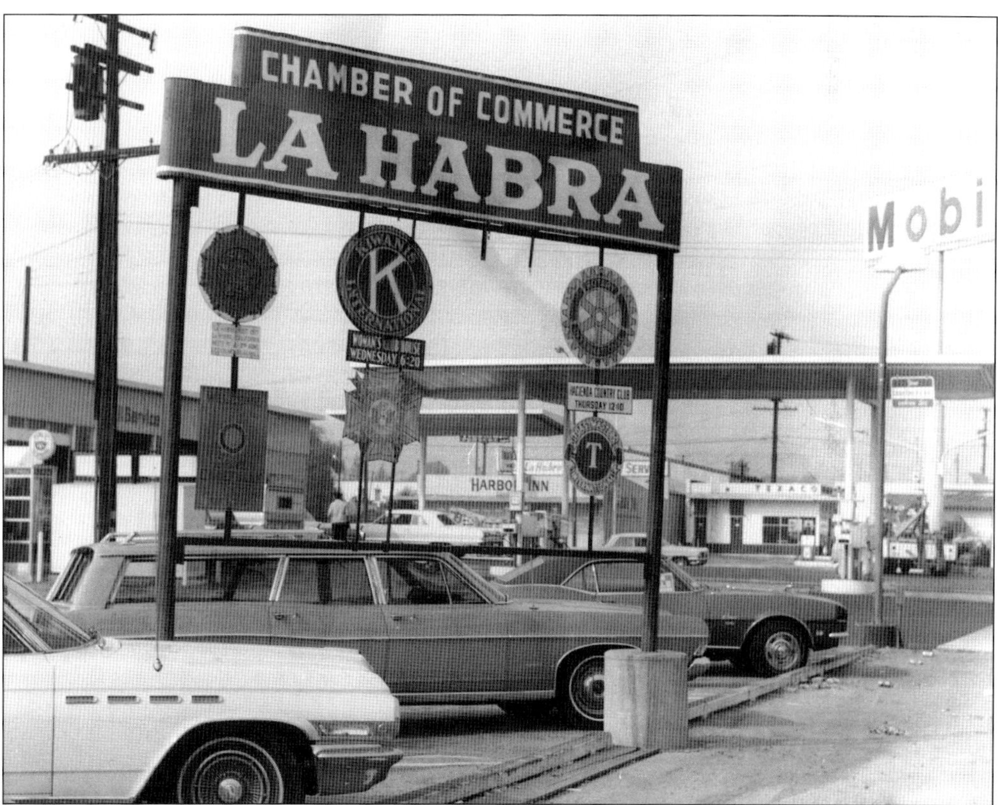

The La Habra Area Chamber of Commerce officially started in 1920, but it has a rich history that dates back to the promotion of agriculture in 1914, when F.R. Aldrich represented the unincorporated town at the Orange County Associated Chamber of Commerce. Adapting to the business evolution of La Habra, the chamber has constantly reinvented itself to promote the current interests of its members, who are often made up of local business owners. This photograph shows the street view of the La Habra Chamber of Commerce in the 1960s. (Courtesy of the La Habra Historical Society.)

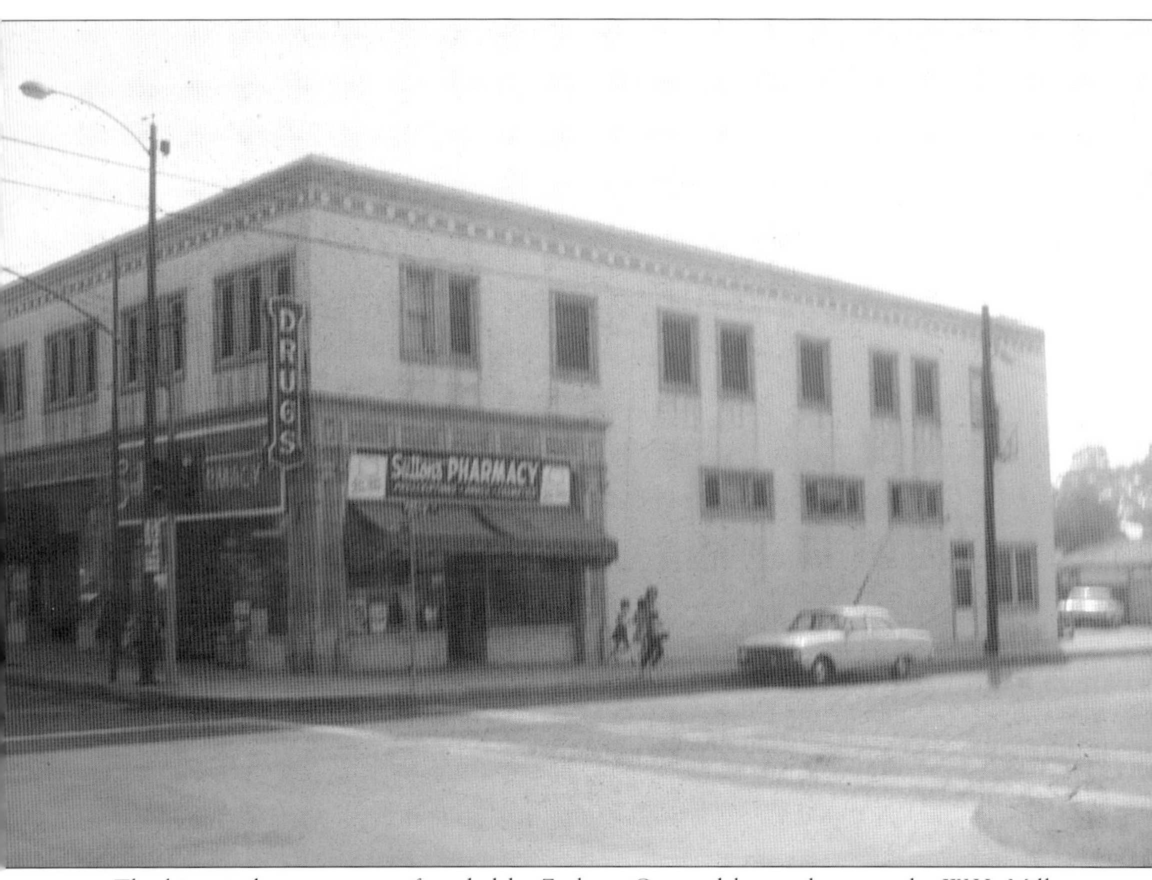

The historical grocery store founded by Zachary Coy and later taken over by W.H. Mills was eventually torn down and replaced in 1922 with, at the time, the most expensive building in the city, costing $50,000 then and nearly $1 million in today's money. The marble and brick building later housed Sutton's Pharmacy, owned and operated by Lyall Sutton from approximately 1944 to the late 1960s. Raised in the city, Sutton worked in the pharmacy during high school under its previous owner and served as a pharmacist after his graduation from the University of Southern California. The drugstore was on the ground floor, with a soda fountain in the front and a prescription department in the back. A physician's office resided downstairs, while several office spaces and boarding rooms took residency upstairs. (Courtesy of the La Habra Historical Society.)

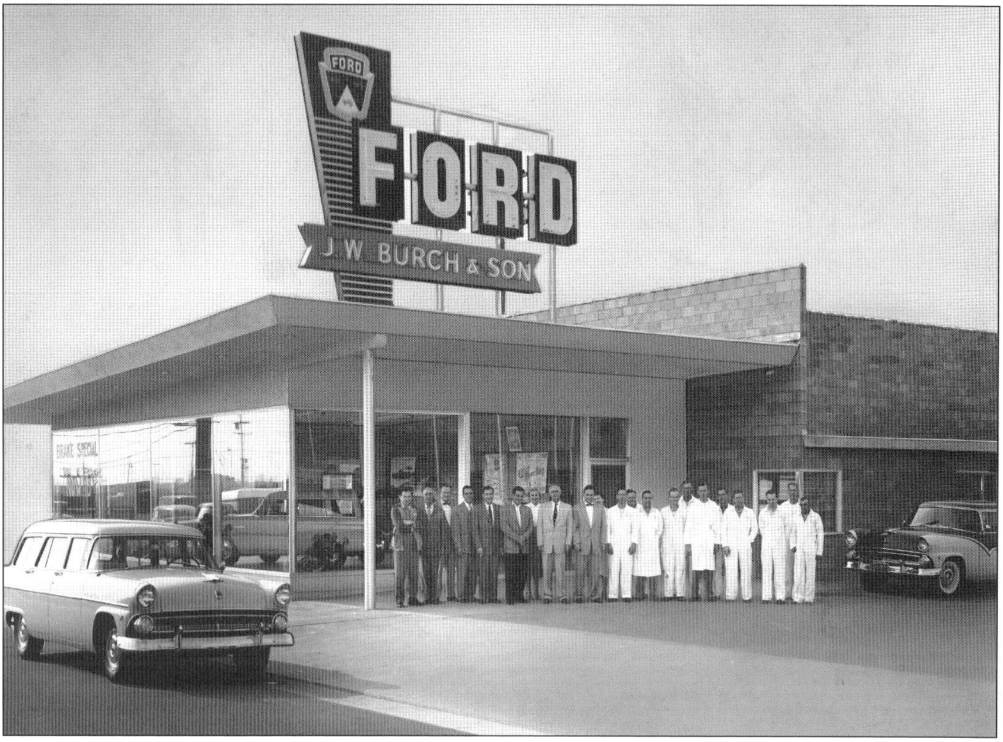

La Habra Burch Ford was owned and operated by the Burch family and opened for business on the corner of Harbor and La Habra Boulevards in the 1930s. The dealership spanned approximately 3.6 acres, which housed both a showroom and an auto shop. The business remained a cornerstone of the community until its closing in 2009. Above, employees are seen proudly posing in front of the dealership, while the photograph below shows the inside of the showroom. The Burch family is responsible for another iconic family-owned and -operated business in La Habra, as they also founded, own, and currently manage La Habra 300 Bowl. (Both, courtesy of the La Habra Historical Society.)

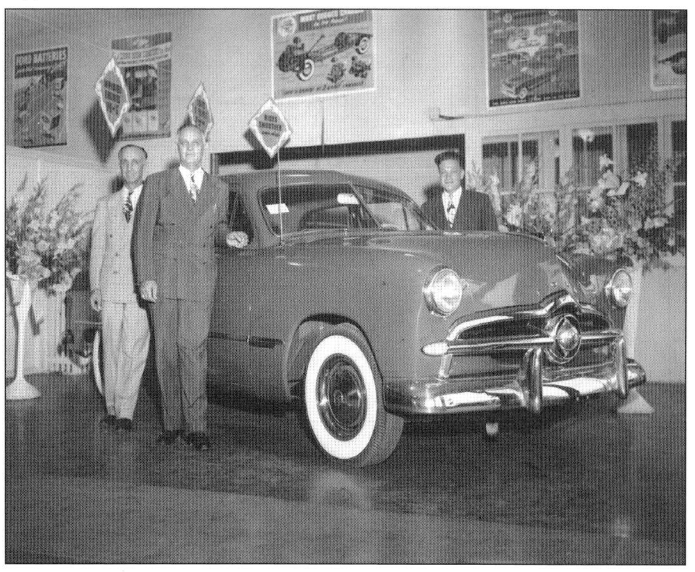

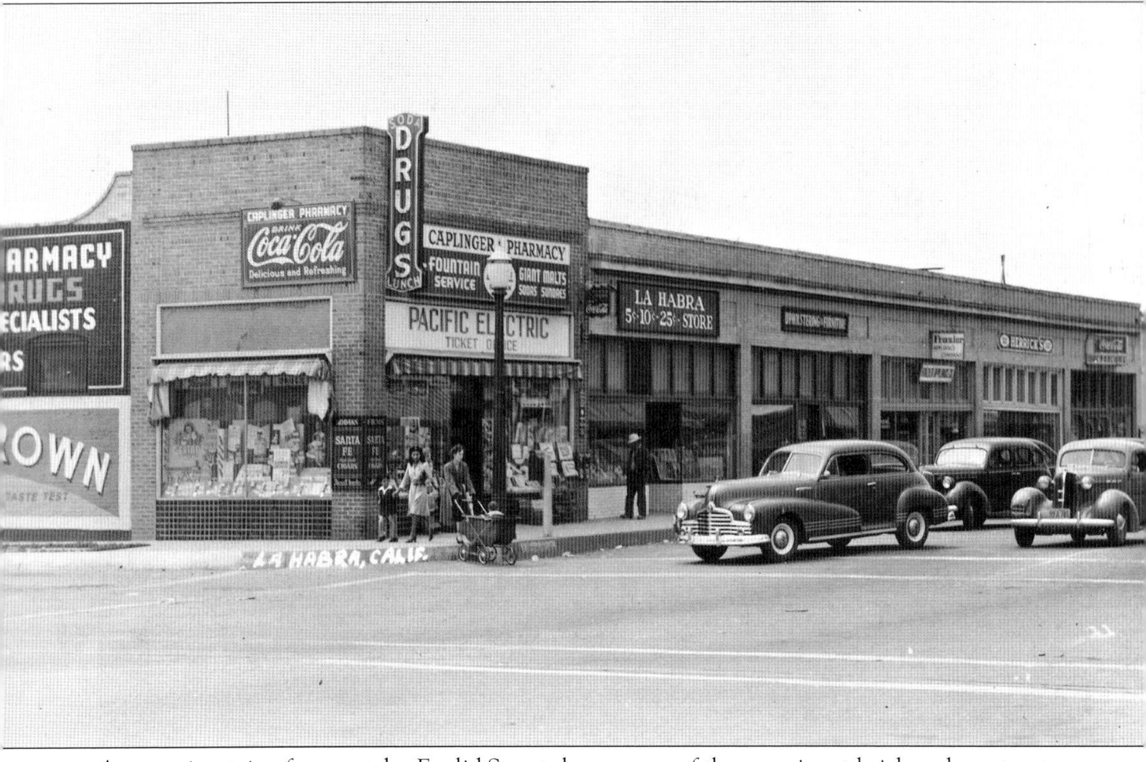

An exterior strip of current-day Euclid Street shows some of the prominent brick-and-mortar stores of La Habra in the 1940s. Caplinger Pharmacy offered residents of La Habra much more than a standard pharmacy; the front also served sodas and ice creams in addition to acting as a ticket office for the Pacific Electric Railway Company. Located off one of the main roads in the city at the time, the store shared the street with many local businesses, providing the citizens of La Habra a walkable shopping experience for a variety of needs. (Courtesy of the Orange County Archives.)

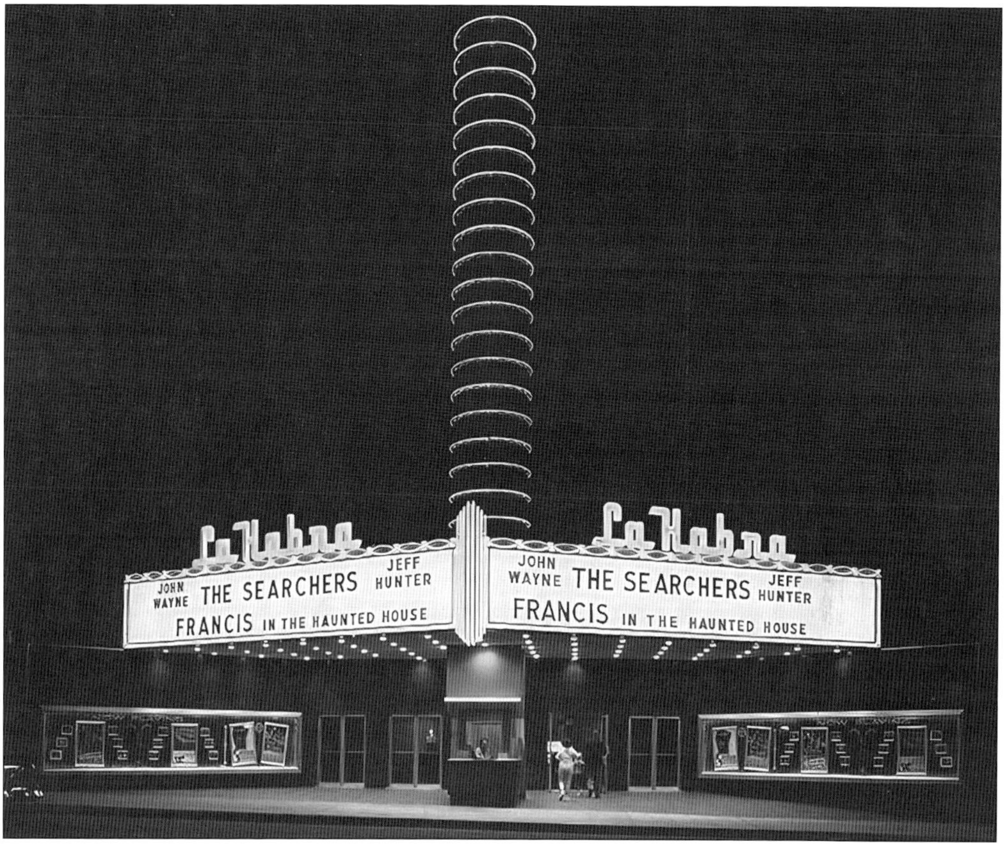

The La Habra Theatre opened in September 1956 with construction completed by Robert L. Lippert Theaters at a cost of $350,000, over $4 million today. On opening night, audiences filled the 1,200-seat, two-screen theater to watch their choice of *Johnny Concho* or *The Man Who Knew Too Much*. During its run, the theater saw several owners before its demolition in the 1990s. (Courtesy of Elana Samson.)

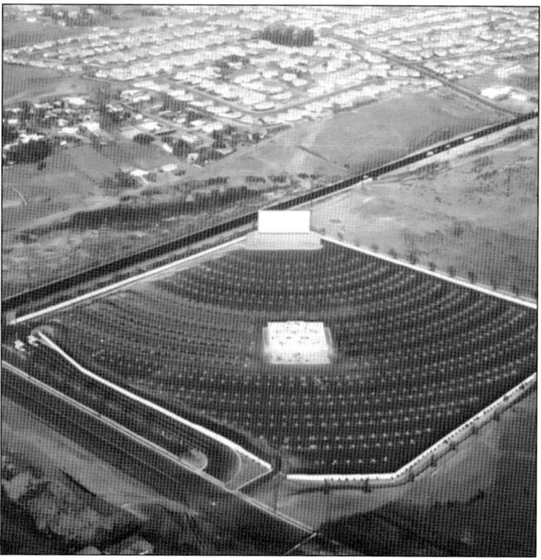

The La Habra Drive-In opened in 1962 with *Boys' Night Out* and *Ride the High Country*. The 1,450-car drive-in would remain in business until its closing in 1989 with showings of *Parenthood* and *Twins*. At opening, it was Orange County's eighth drive-in theater; with its closing, the county was left with five, signaling the end of this era of entertainment. (Courtesy of the Los Angeles Public Library.)

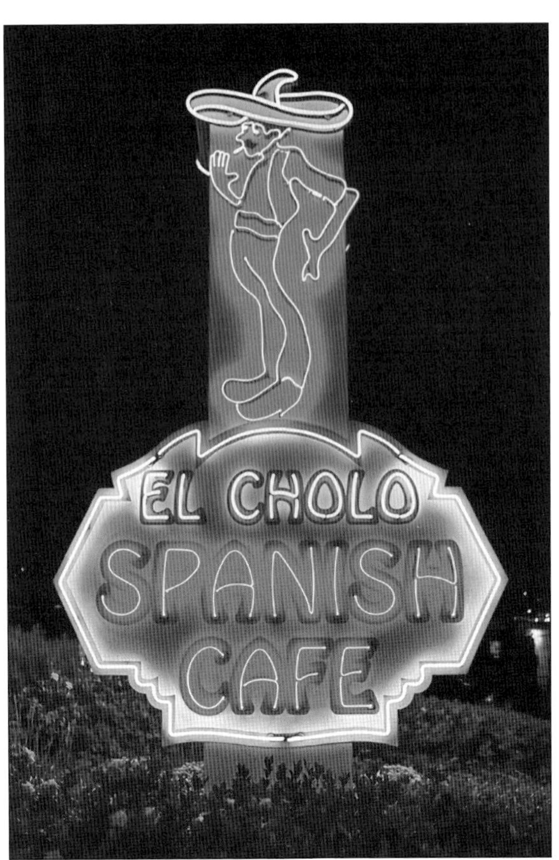

Founded in Los Angeles in 1923 under the name "Sonora Café," El Cholo opened a second location in La Habra in 1962. The owners changed the name after a customer drew a caricature called "El cholo"—referring to the field hands of the Spanish settlers—and adopted the drawing as their mascot. The restaurant has flourished in La Habra for over 60 years. (Courtesy of the City of La Habra.)

Before it was the Chicken Box, another chicken restaurant known as Dixie Chicken occupied the same building that the longtime La Habra restaurant occupied for several decades. Keen eyes can see that the Chicken Box retained the same chicken icon for its restaurant when it took over the space next to La Habra 300 Bowl. (Courtesy of the La Habra Historical Society.)

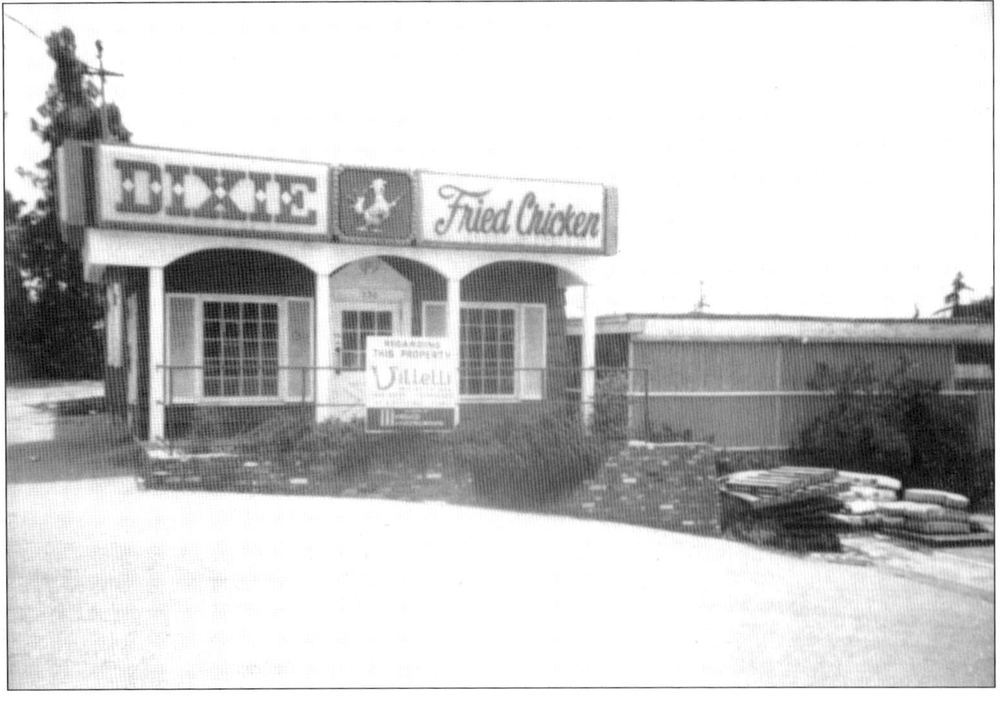

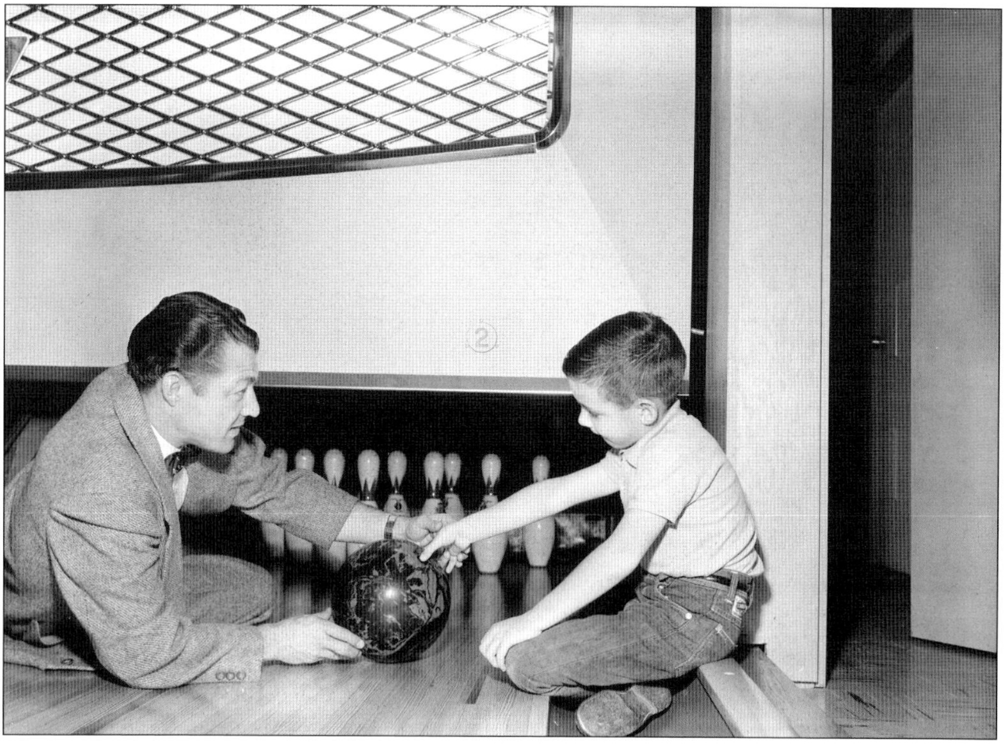

Former La Habra 300 Bowl manager Bob Allison poses with a young bowler in December 1960 on one of the alley's numerous lanes. This heartwarming picture was taken during one of the many fundraisers held across Orange County during that time period to raise funds for the fight against polio. Since its opening in 1958, La Habra 300 Bowl has remained a Burch family–owned and –operated business and is instantly recognizable by its Googie architectural style, including the A-frame roof, pylon beams, and its iconic neon starburst sign, pictured below. (Above, courtesy of Orange County Archives; below, courtesy of La Habra 300 Bowl.)

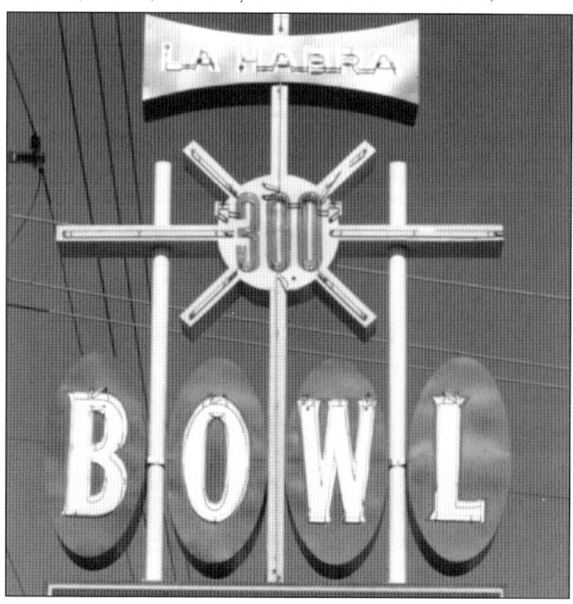

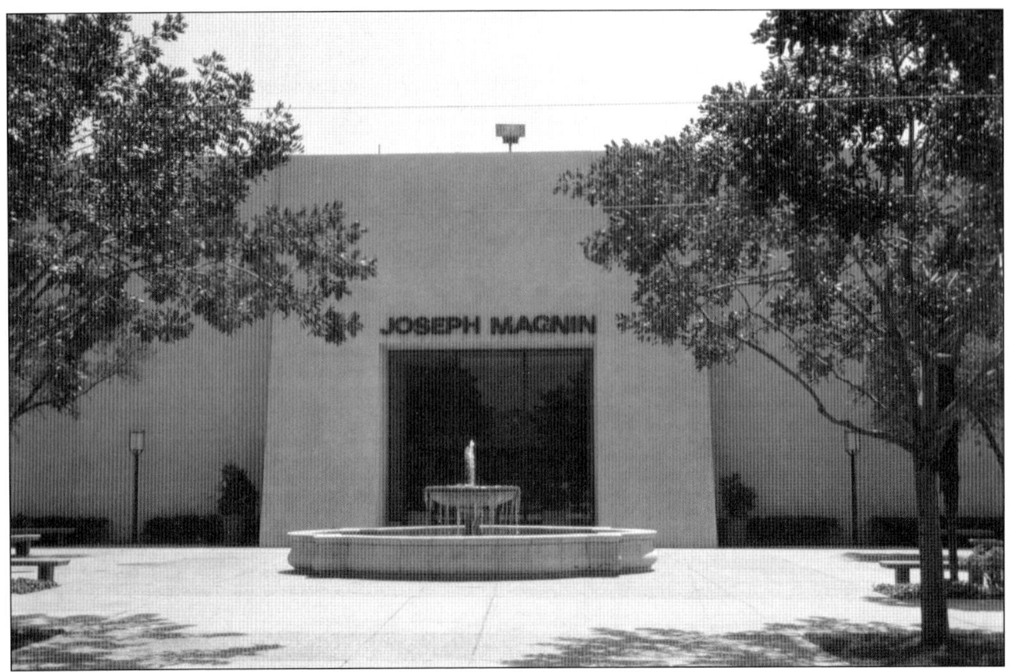

Joseph Magnin (pictured above in May 1975) and Buffums were among some of the prominent department stores that were included in the grand opening of La Habra Fashion Square on August 10, 1968. At 19,500 square feet, this community shopping center was the largest and final "Fashion Square" mall built by the Bullock's department store chain. The La Habra Fashion Square closed under 30 years of operating in 1992 and has since been redeveloped to house the La Habra Marketplace, where stores such as Smart & Final, Regal Cinemas, and Sprouts currently operate. (Both, courtesy of Orange County Archives.)

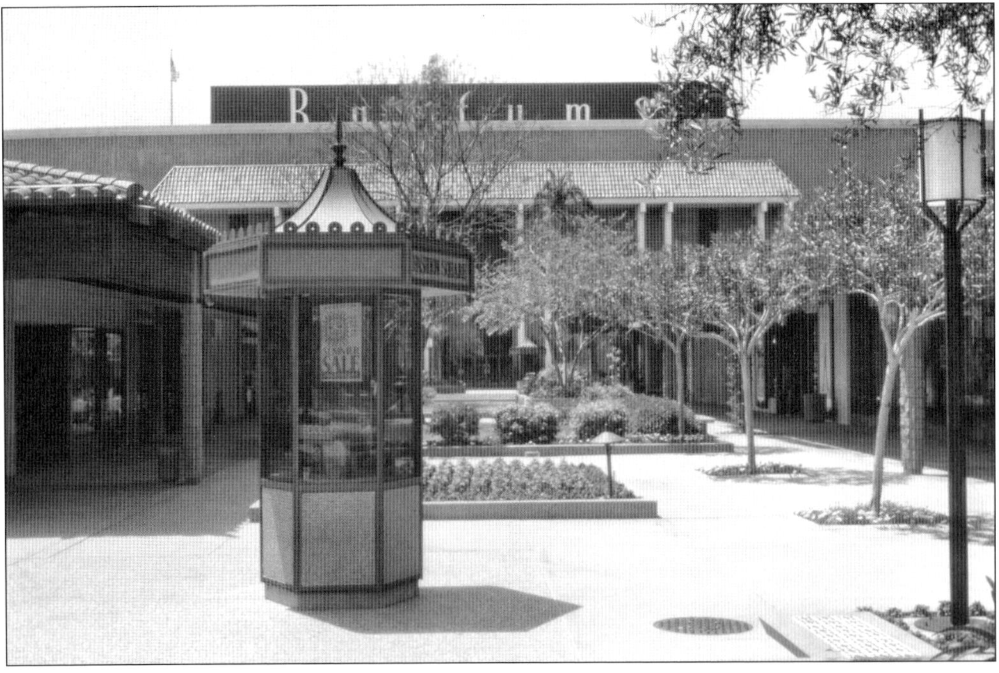

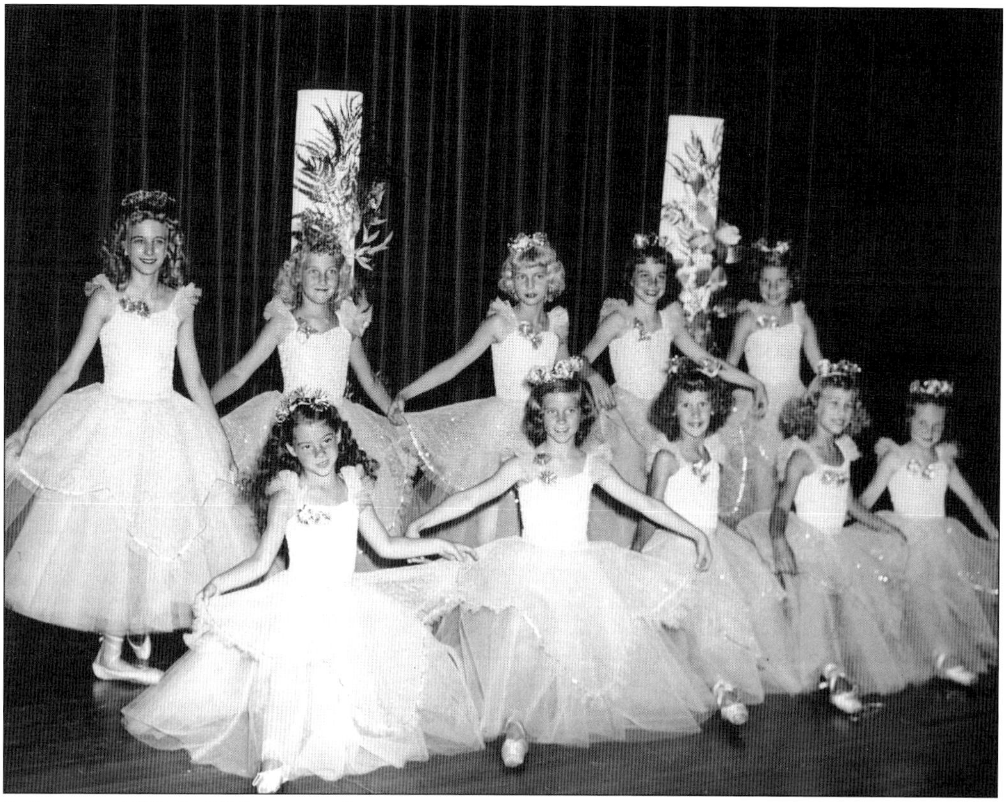

In January 1953, Jo Walin opened Jo's Dance Studio at 801 Whittier Boulevard. For over 20 years, Walin taught generations of students dance forms such as ballet, tap, and more. A Hollywood-trained dancer herself, Jo Walin danced in the 1938 film *Happy Landing*. (Courtesy of the City of La Habra.)

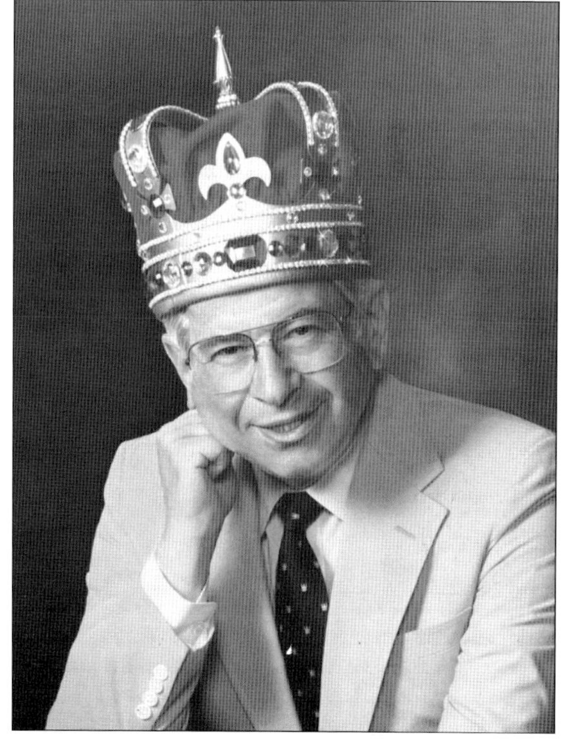

Paul Goldenberg, "King of Big Screens," started a small TV-repair shop in Los Angeles before building his La Habra business into one of the largest single-store television retailers in the country for nearly 40 years. Yet, according to Goldenberg, his greatest accomplishment was his philanthropy, giving to the City of Hope, the Los Angeles Jewish Home, and scholarships for high school students. (Courtesy of the Boys & Girls Club of La Habra.)

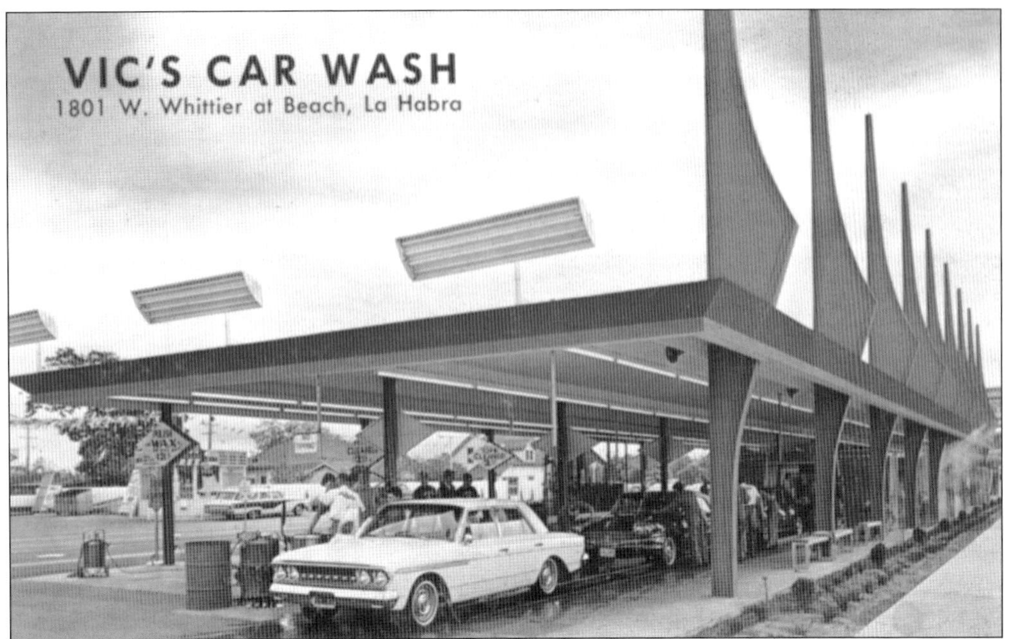

Located just west of the site of the current Buff and Shine Car Wash, Vic's Car Wash kept La Habra's vehicles squeaky clean largely by hand washing techniques, although there was some automation during this time, as seen upon a closer look at this postcard of the conveyor belt and water misters. (Courtesy of the Thomas Pulley collection.)

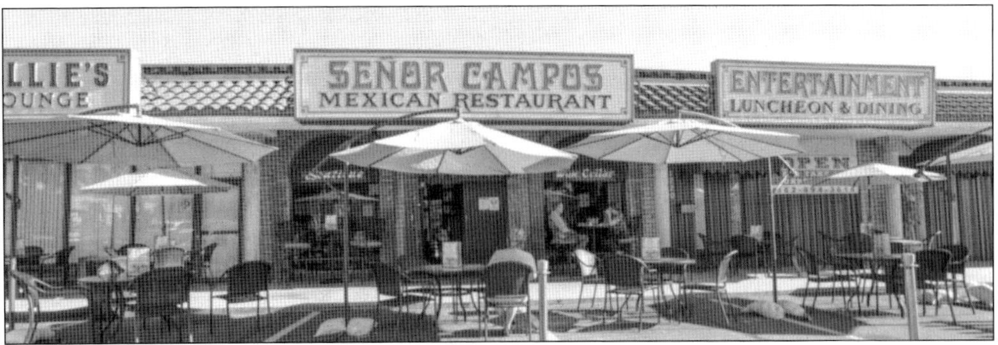

Señor Campos was founded in 1974 by John and Tillie Campos to create a restaurant that would make patrons feel loved and taken care of. They operated the restaurant for many years, and then, when they retired, they sold it to longtime staff. For over 30 years, Señor Campos has hosted children from the La Habra Boys & Girls Club for a Christmas party every year. (Courtesy of the City of La Habra.)

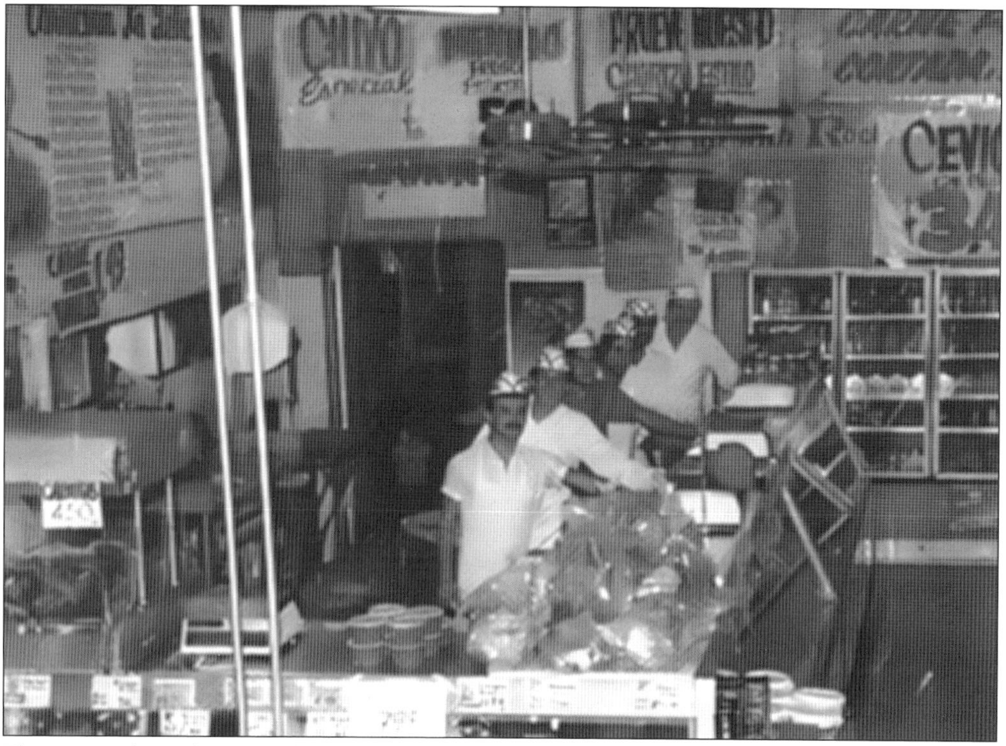

The roots of Northgate Market trace back to the González Reynoso family, who immigrated to the United States from Jalisco, Mexico, in the early 1970s. They settled in Anaheim, California, in 1980 when they purchased their first small market, where the González family could share their Mexican heritage's authentic tastes and flavors with the local community. In 1986, the González family chose La Habra for their second location, where many siblings, cousins, and uncles worked. Despite its growth, Northgate Market has remained a family-run business, with members of the González Reynoso family actively involved in its operations. (Both, courtesy of the City of La Habra.)

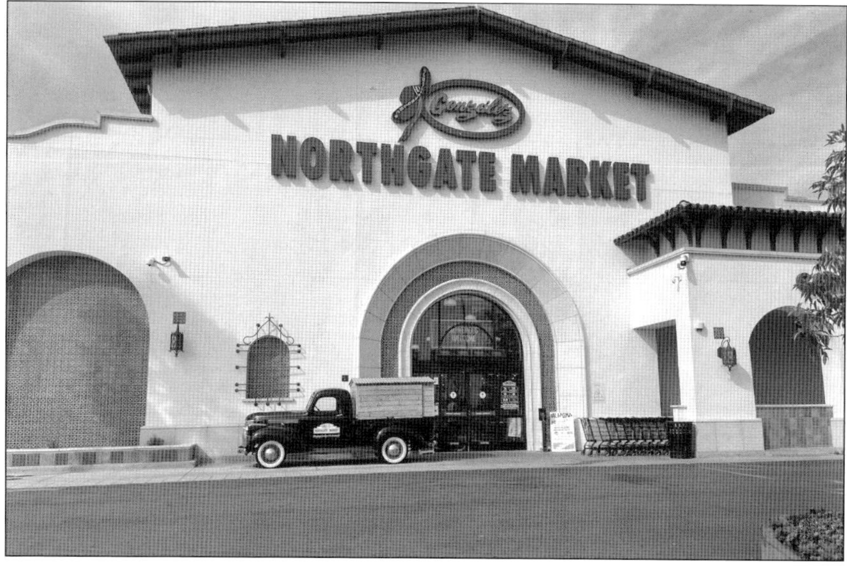

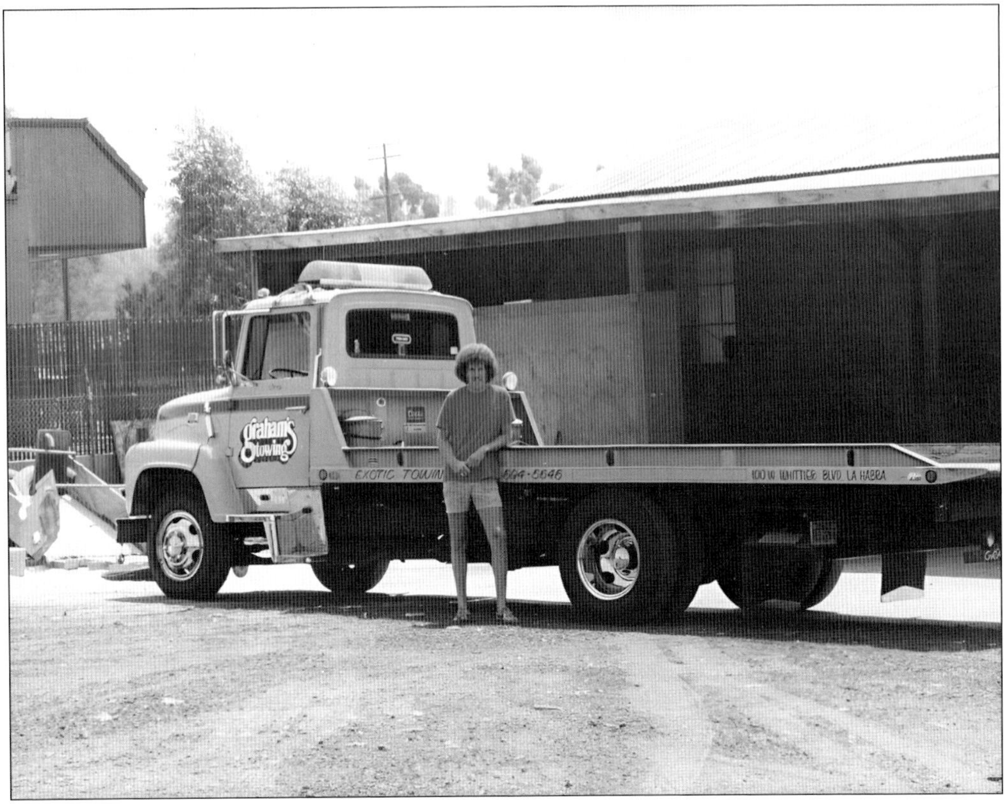

This picture was taken in 1979, shortly after Danny, Judy, and Tom Hanson purchased Graham's Towing Service from John and Bill Graham, who also owned the Chevron station on Euclid Street and Whittier Boulevard. The tow service was moved to another property in 1985 and again in 1999 to its current address of 1501 Lambert Road. (Courtesy of Ofelia Hanson.)

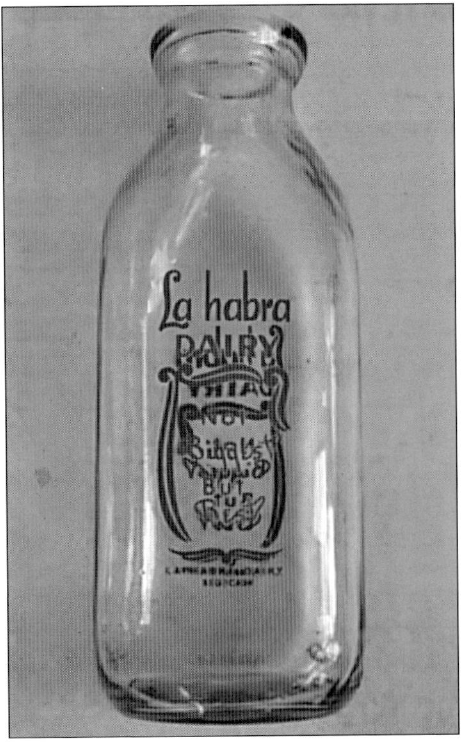

This milk bottle from the La Habra Dairy reads, "Not Biggest but Best," an apt slogan for the dairy located on Harbor Boulevard and Lambert Road. It became Orange County's last roadside dairy with cows on site in 1976. La Habra Dairy was bought in 1954 by Little Home Dairy, which brought its own plucky slogan, "You can whip our cream, but you can't beat our dairy!" (Courtesy of Elana Samson.)

Six

A Growing Suburban Community

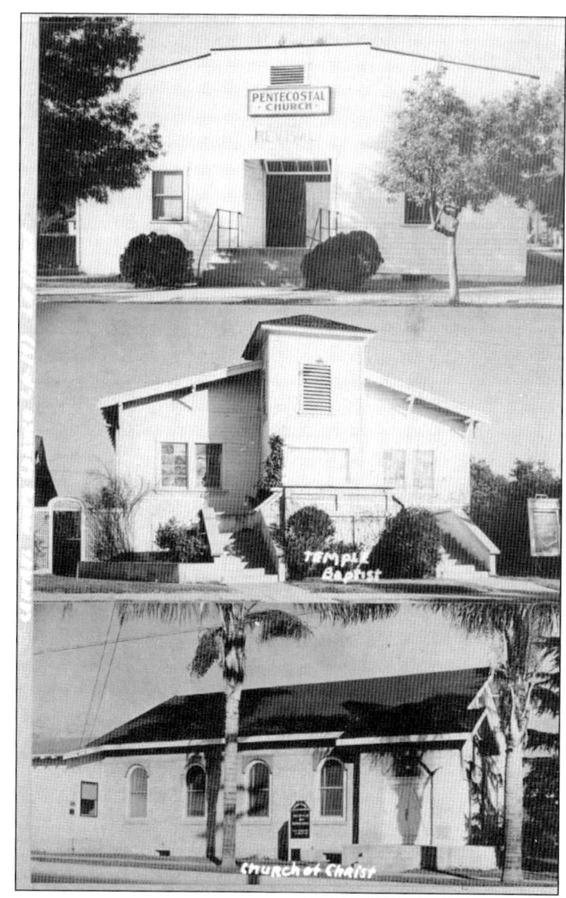

At the top of this postcard is the Pentecostal church, with the Temple Baptist Church in the middle, built around 1922 on North Hiatt Street. Last is the Church of Christ building on Cypress and First Avenue, which was dedicated in 1923 after the group's first meetings were held in a tent. Today, the congregation goes by the La Habra Christian Church. (Courtesy the Thomas Pulley collection.)

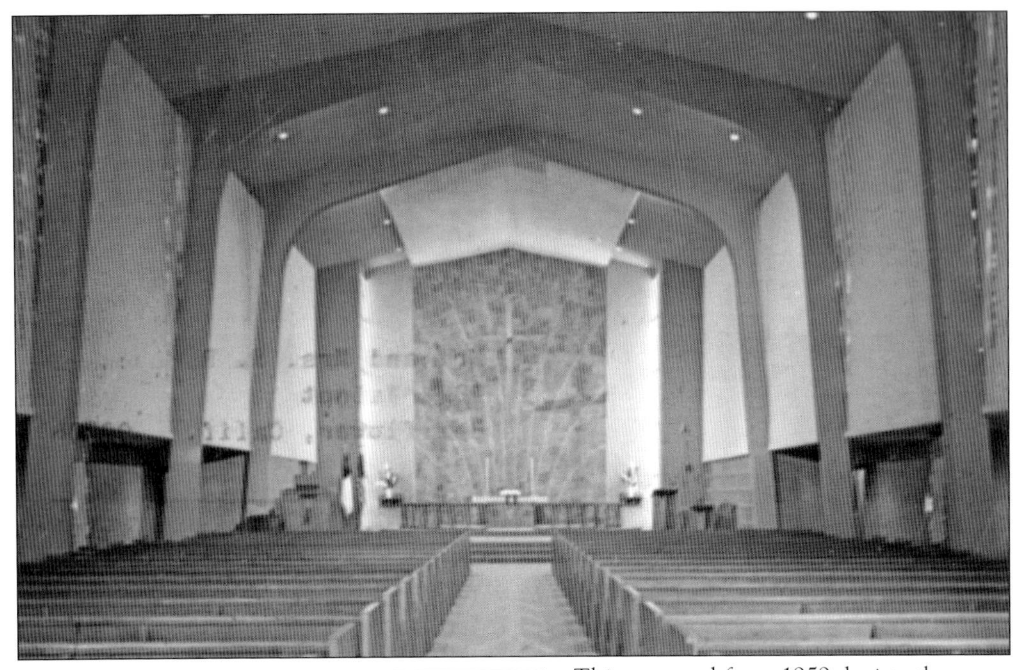

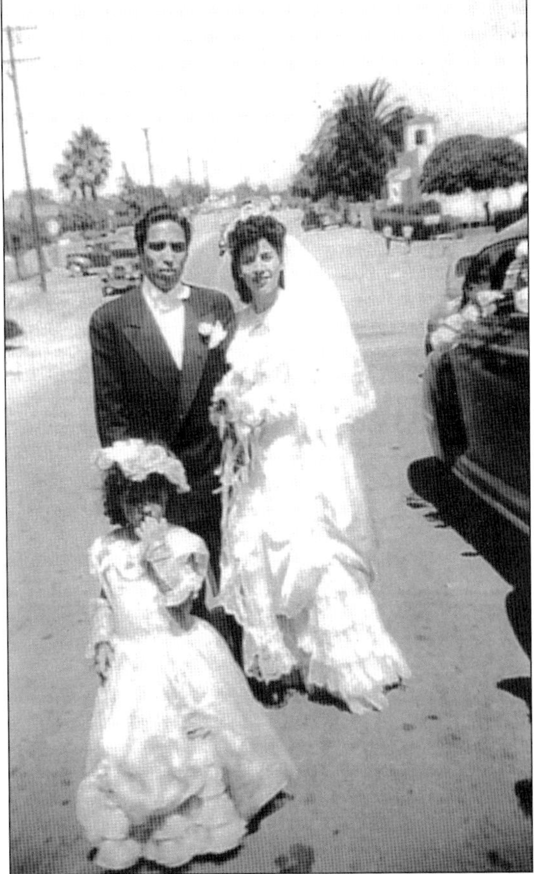

This postcard from 1959 depicts the sanctuary of the La Habra United Methodist Church, which is celebrating 125 years in La Habra in 2025. The back of this postcard has the cheeky hand-written message of "As beautiful as our sanctuary is, it looks twice as good when YOU are here." (Courtesy the Thomas Pulley collection.)

With some previous community engagement in 1906, the Baptist church became active again in the city in 1921. The Temple Baptist Church had 100 members by 1922 and erected its first building with volunteer help in the same year. It moved to a new site the following year and remodeled with additions. Here, a young couple stands outside the Temple Baptist Church following their wedding. (Courtesy of Tony Gonzales.)

The photograph at right depicts the second Our Lady of Guadalupe Church Building. The first structure was destroyed in a fire in 1944, but the church was rebuilt on the same site just three years later in 1947. At this point, the church building was located near the Fourth Street, or Alta Vista, migrant camp. After the congregation moved to La Habra Boulevard, it became first the Gary Center and then the La Habra Boxing Club before it closed its doors in 2020. Below is a photograph of the construction of the current Our Lady of Guadalupe Church, which was dedicated in 1971. (Both, courtesy of the La Habra Historical Society.)

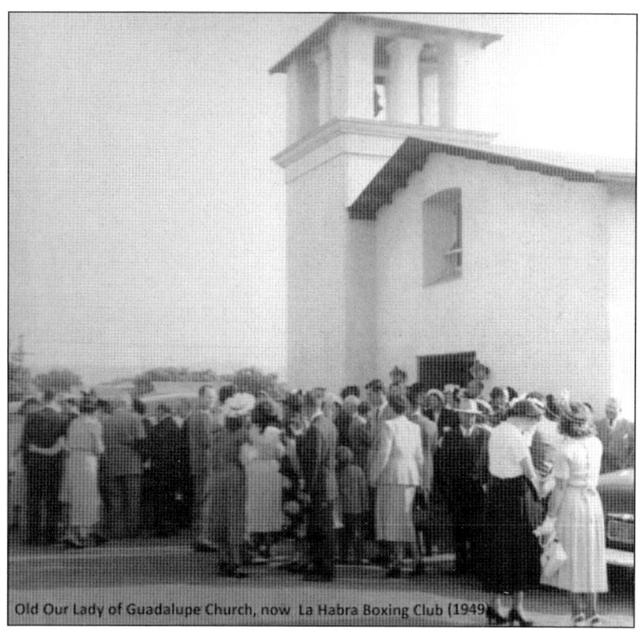

Old Our Lady of Guadalupe Church, now La Habra Boxing Club (1949)

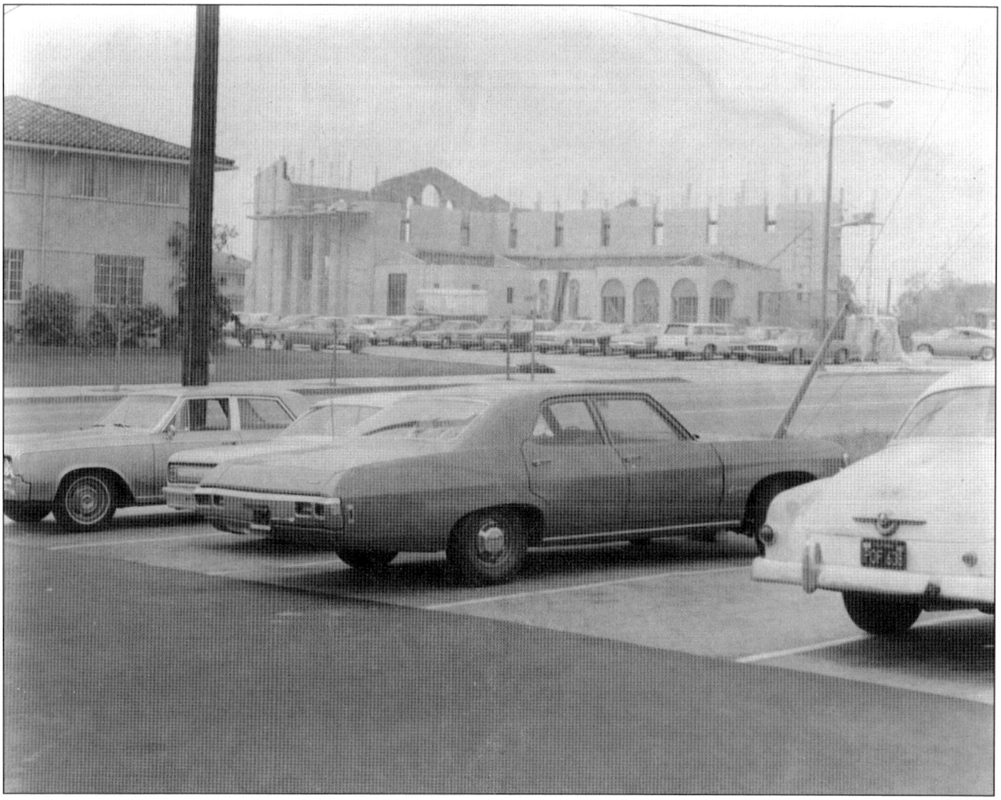

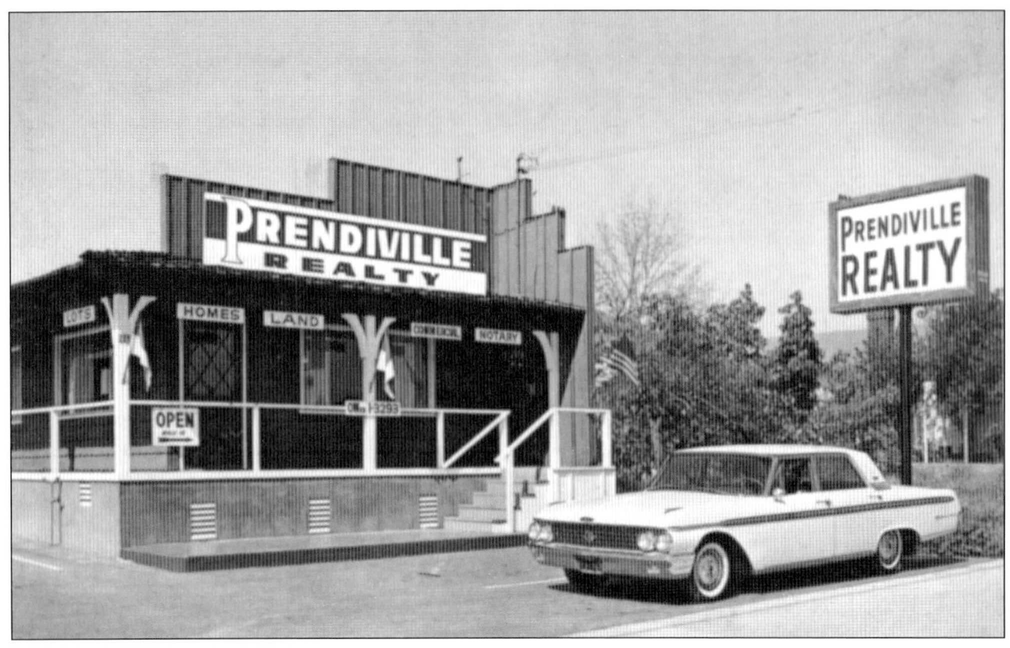

Prendiville Realty operated in La Habra during the 1960s. The founder and owner, Garry Prendiville, also served as president of the La Habra Area Board of Realtors from 1962 to 1963 and as vice chairman of the city planning commission. Operating as a realtor during this time in La Habra would prove to be a lucrative decision, as the population experienced a rapid growth. Between 1950 and 1960, La Habra's population increased from 4,961 residents to 25,136 residents, over a 400 percent jump. In the next decade, the population would grow by an additional 65 percent to 41,350. Many houses, like the one photographed here, were built to meet the demand. (Both, courtesy of the Thomas Pulley collection.)

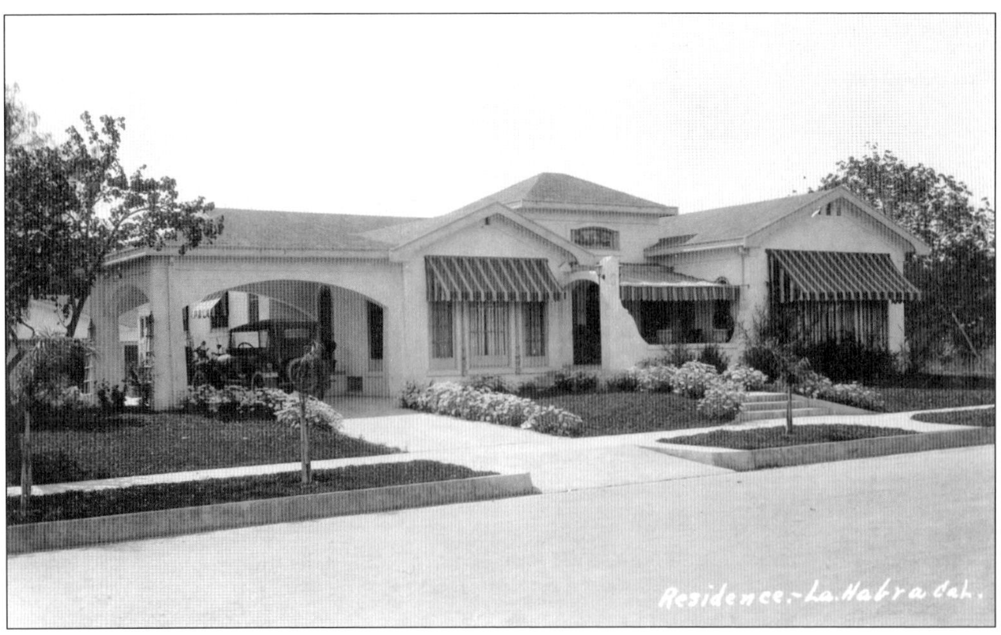

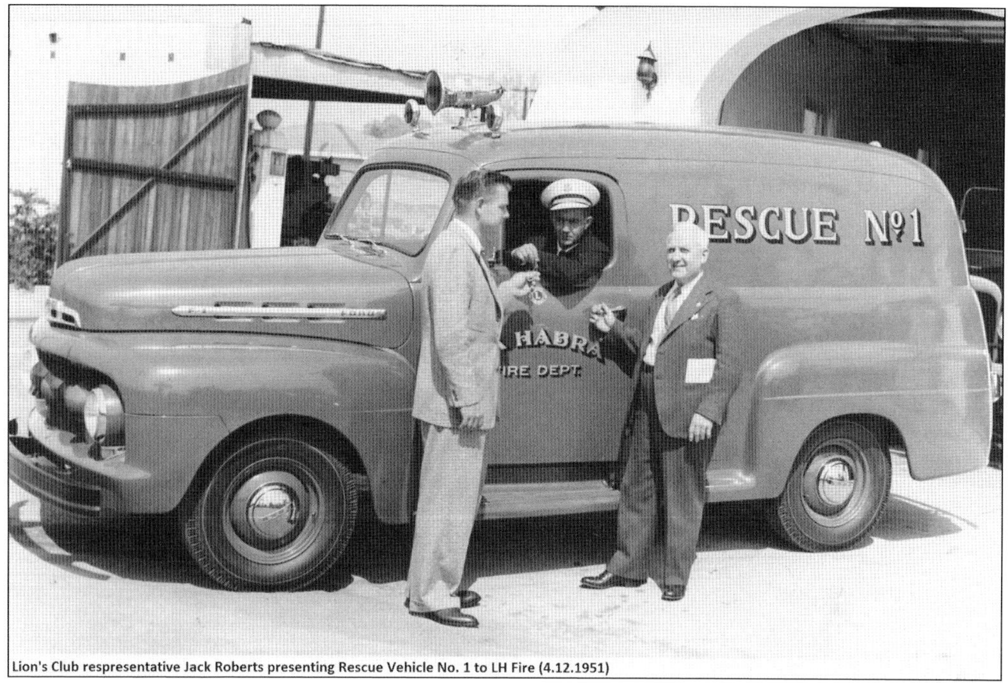

La Habra Lions Club representative Jack Roberts presents Rescue Vehicle No. 1 to the La Habra Fire Department in 1951. Donations and fundraising were important for equipment acquisition for much of the La Habra Fire Department's early history. Dance fundraisers were often held in the first years of the department's establishment. (Courtesy of the City of La Habra.)

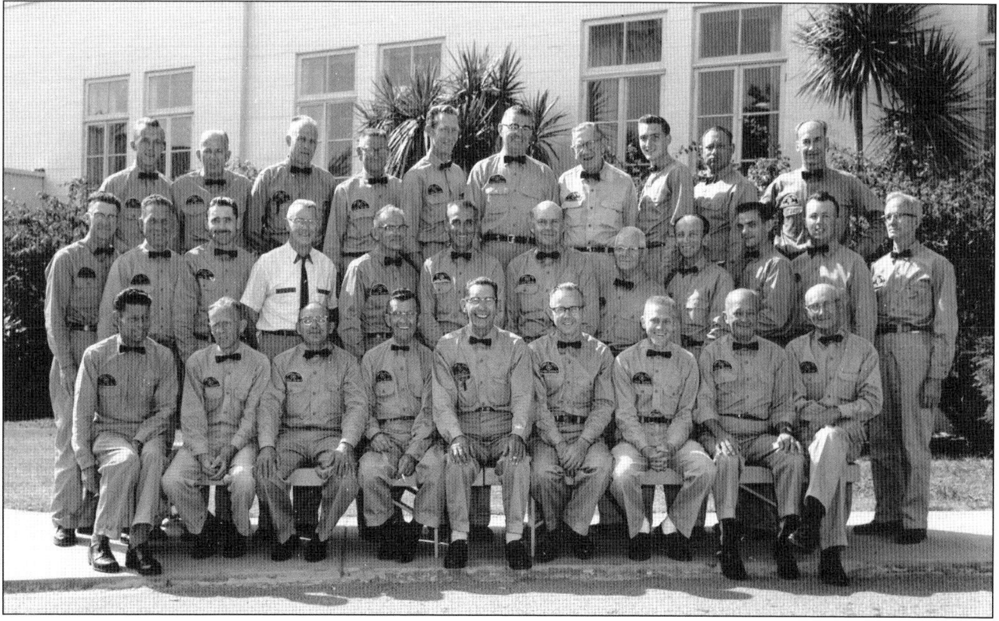

This group photograph of the Lowell Unified School District's bus drivers was taken some time in the 1950s. School bus usage has declined sharply in the last few decades, but at the time of this photograph, the service was integral to the transportation of students to class. (Courtesy of the City of La Habra.)

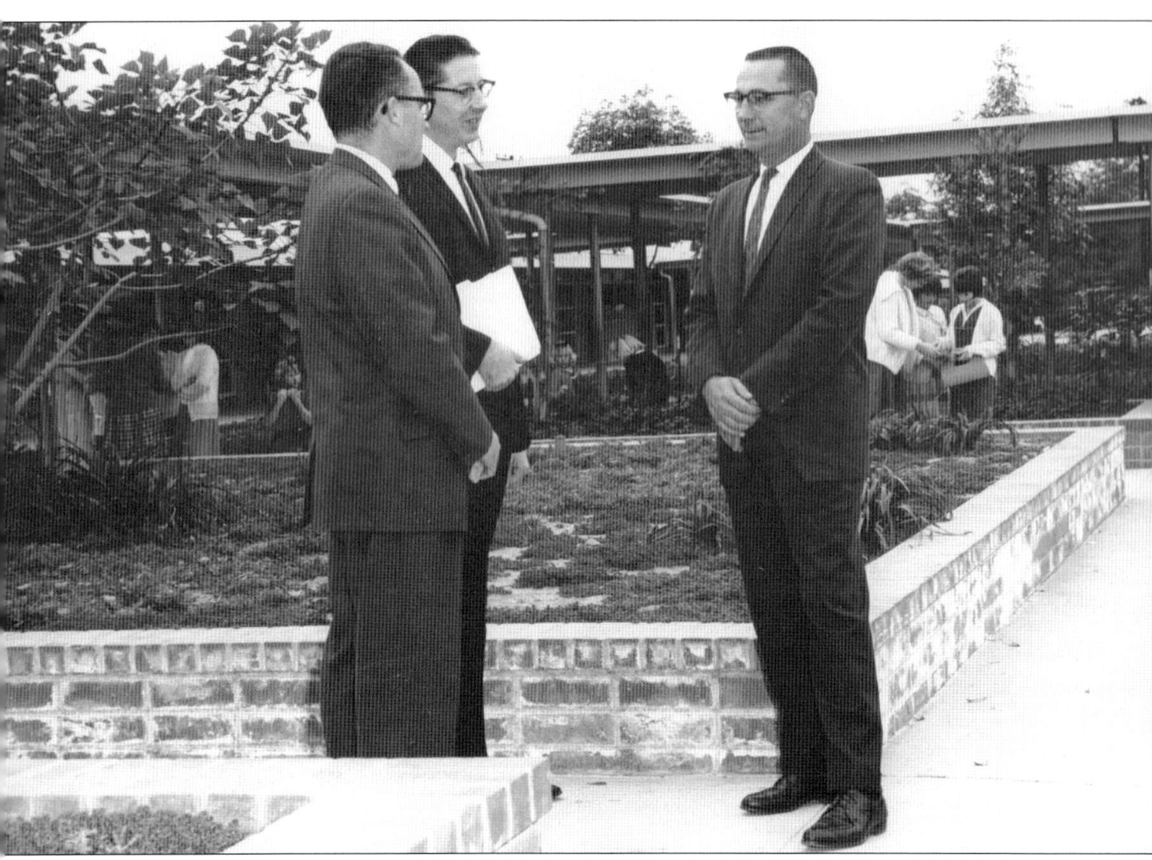

Efforts to establish schools in the Lowell area started in 1906, making it one of the few "joint" (governed by two counties) districts in the state. Parents in the easternmost portions of La Habra grew frustrated once their students would return home after dark due to long commutes on the school bus, so a plan for the long-talked-of school in the Lowell area became a reality. A schoolhouse in Whittier was purchased and relocated to Valley Home Avenue to serve as the first Lowell School, needing telephone wires to be cut and trees hacked to accommodate its transportation. Growing enrollment numbers caused the addition of many new schools in the district, and today, Lowell Joint School District houses many La Habra campuses. (Courtesy of the City of La Habra.)

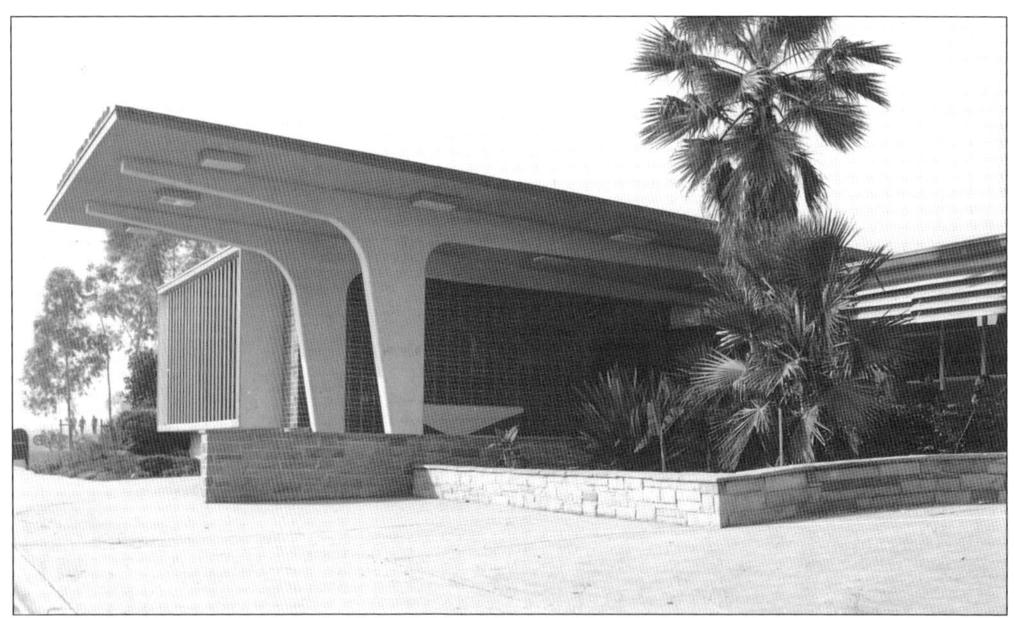

La Habra did not have a high school of its own until 1954, but by 1966, it had three. The first of these was La Habra High School, which appears in this photograph taken in 1956, two years after it opened in 1954, and features the same front facade many are still familiar with. In recent years, La Habra is perhaps best known for its championship football team, which has won seven California Interscholastic Federation (CIF) titles since 2002, the most recent awarded in 2015. The girls' volleyball and men's soccer teams have also won multiple state titles. (Above, courtesy of the La Habra Historical Society; below, courtesy of the *Orange County Register*.)

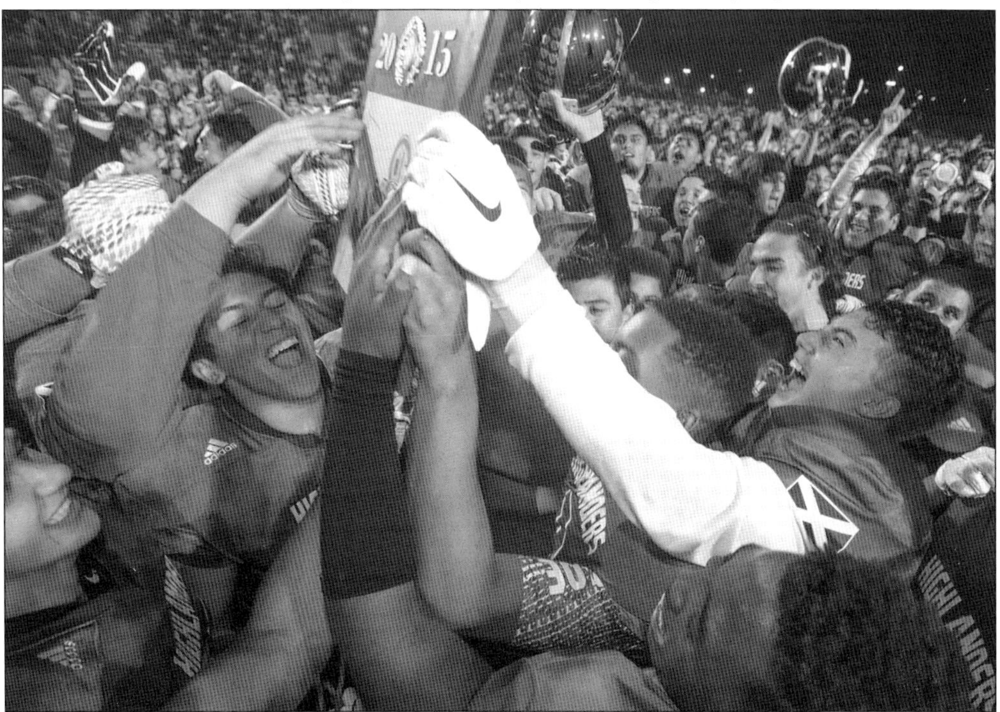

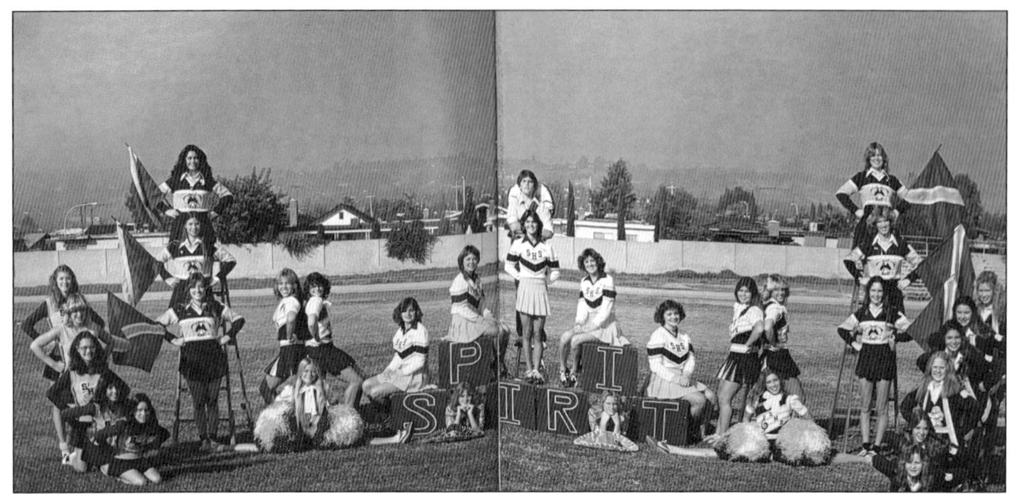

This is a photograph of the pep unit at Sonora High School from its 1979 yearbook. Sonora was the third high school to open in La Habra in 1966 and has been known for its water polo and basketball teams as well as its International Baccalaureate program. Famous former Sonora students include basketball legends Ann and Dave Meyers and *Avatar* and *Titanic* director James Cameron. (Courtesy of the La Habra Historical Society.)

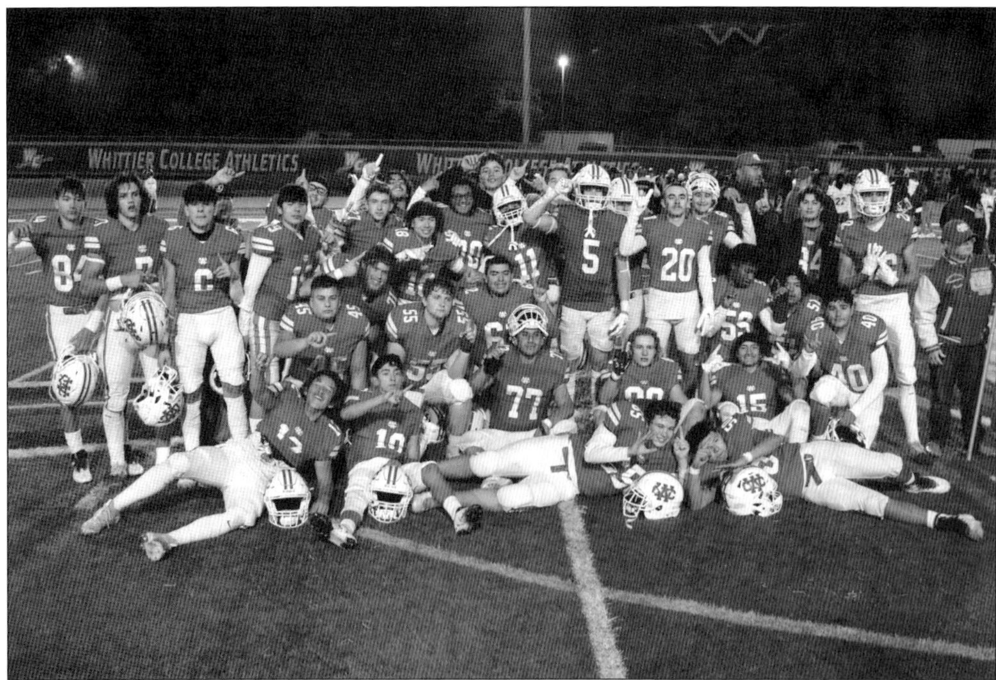

Though Whittier Christian was founded in 1958, it did not move to its present-day location in La Habra until 1981. The nondenominational Christian high school has won 102 league championships in addition to 18 CIF championships since its founding. Pictured here is the school's 2022 state championship football team. (Courtesy of Whittier Christian High School.)

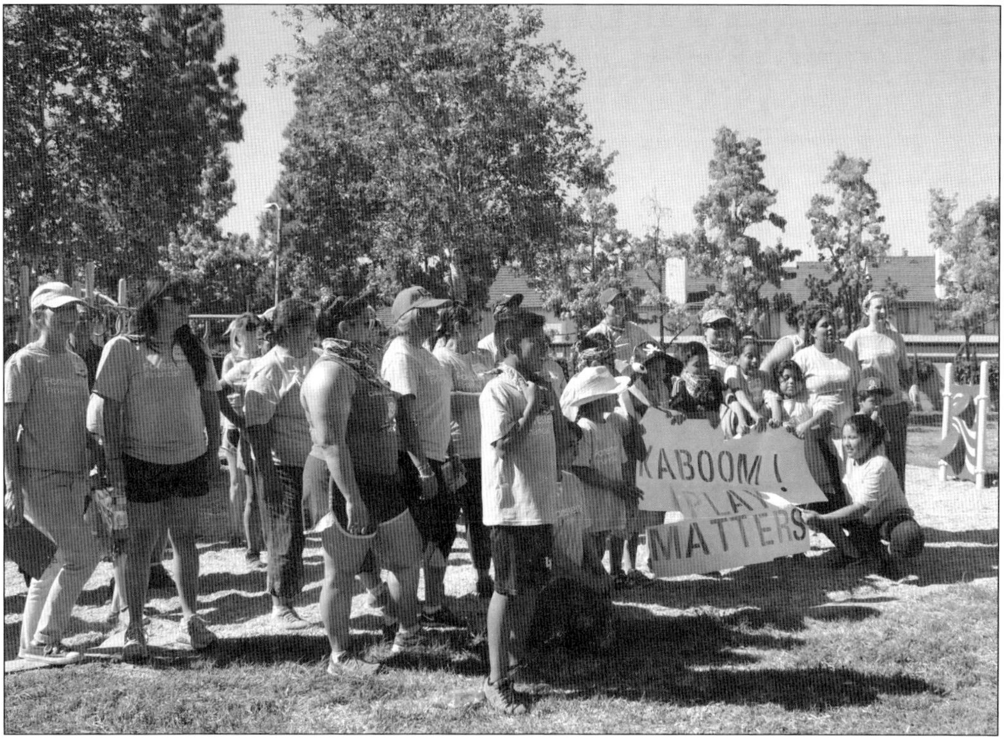

KABOOM! play spaces are built with trust and collaboration with each community. The organization believes in the power of community to unify, heal, and rebuild, and that lies at the heart of its work and mission. KABOOM! has helped the City of La Habra rejuvenate neighborhoods by working with civic leaders, volunteers, and sponsors to renovate eight playgrounds within La Habra, and the city is hoping to partner with it in a new neighborhood again soon. Pictured above is a group shot at the Montwood Park KABOOM!, and below is an aerial shot taken during the Portola Park KABOOM! renovation. (Both, courtesy of the City of La Habra.)

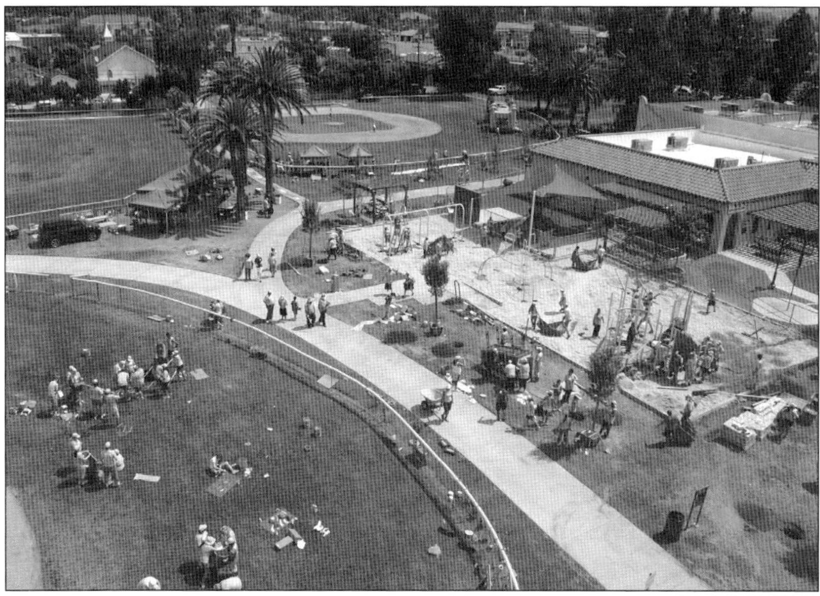

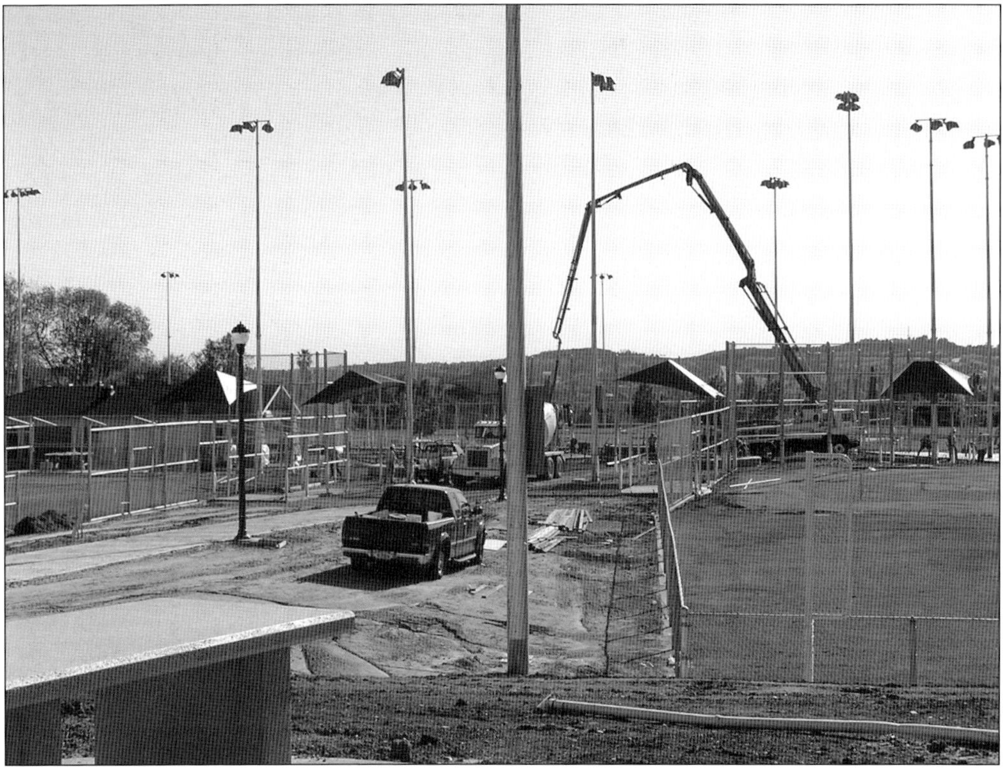

The City of La Habra purchased the 22.6 acres of La Bonita Park in 1957, and development was initiated in 1960. Prior to the renovation project, which began in 2002, the park only had one softball field. Today, the park boasts four lighted softball fields, a cobblestone concession shack, a playground, and a winding walking path. Pictured above is a snapshot taken during the construction of the softball field complex, and below, city council members and officials throw out the first pitch during the first opening day ceremony for the La Habra Girls Softball Association's spring league. (Both, courtesy of the City of La Habra.)

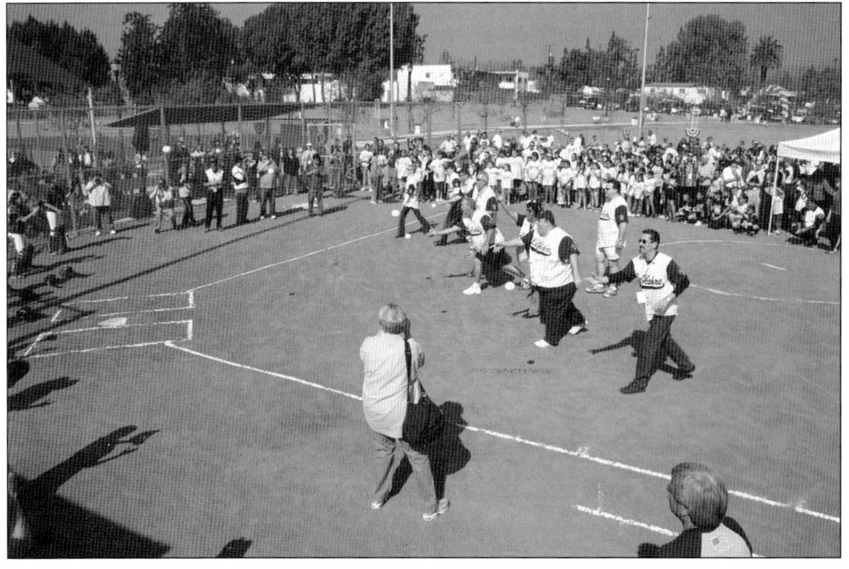

Seven

City Celebrations through the Years

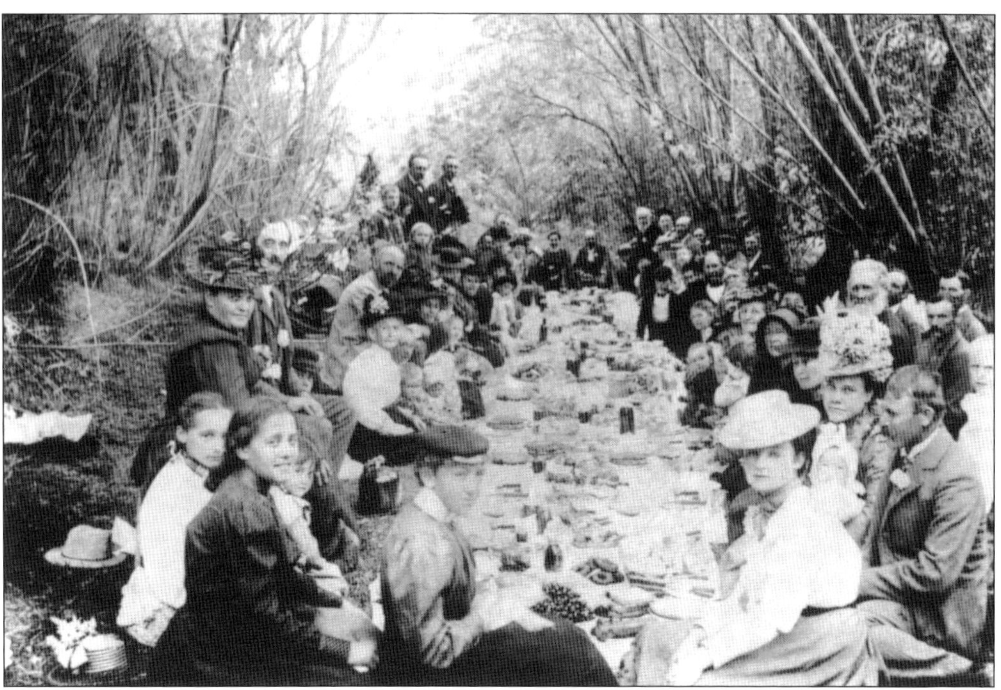

Founding members of La Habra, including "father of La Habra" Willets J. Hole (standing at center rear) and early Swiss settlers Luehm (left front) and Leutwiler (right front) families, attend the second annual Old Settlers' Picnic. Hosted by the Ladies' Mutual Improvement Club of La Habra, the picnic took place at Martinez Grove in 1899 and evokes the same feeling of community that is still a cornerstone of life in La Habra. (Courtesy of the La Habra Historical Society.)

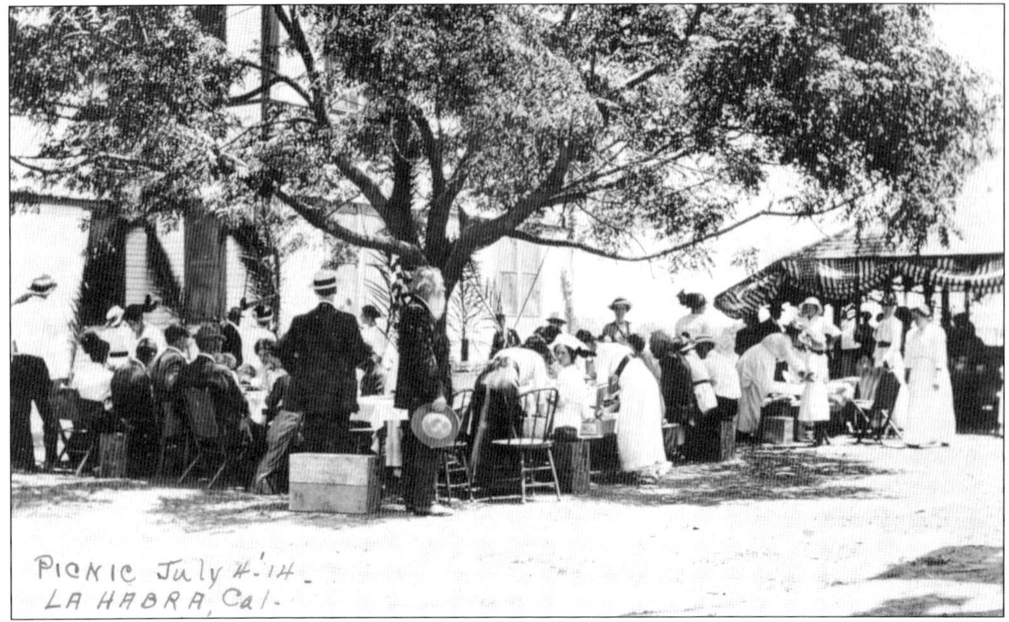

A highly anticipated holiday celebration in La Habra, this picture shows a Fourth of July picnic in 1914. While on a much smaller scale than the Fourth of July celebrations of today, which typically include pyrotechnic shows of around 400 individual fireworks, community members still gathered with their friends and family over food and camaraderie to observe the holiday. (Courtesy of the Thomas Pulley collection.)

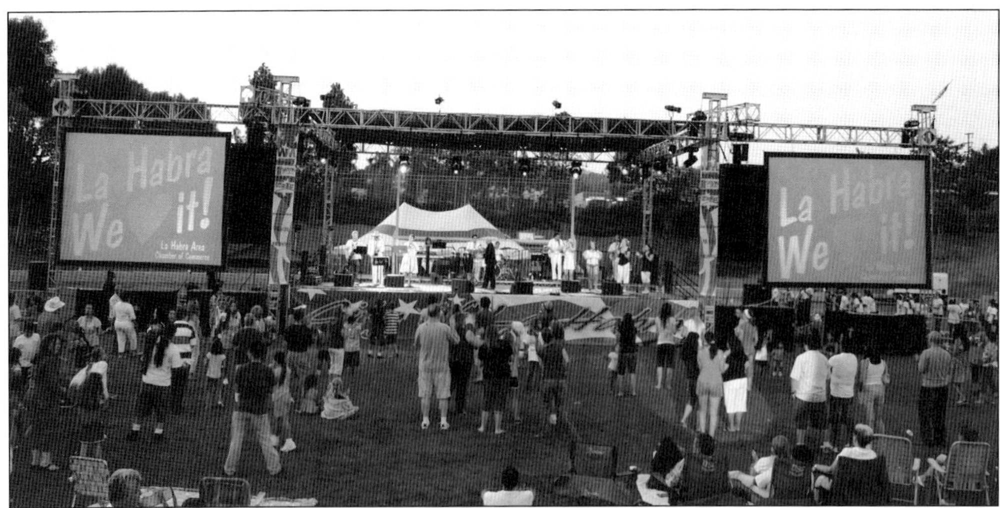

As the sun goes down, the crowd starts to grow around the stage in this photograph from the 2007 Fourth of July celebration. It is estimated that around 10,000 spectators attended the event each year while it was held at La Habra High until it moved just down the street to La Bonita Park to accommodate growing crowds in 2018. (Courtesy of the City of La Habra.)

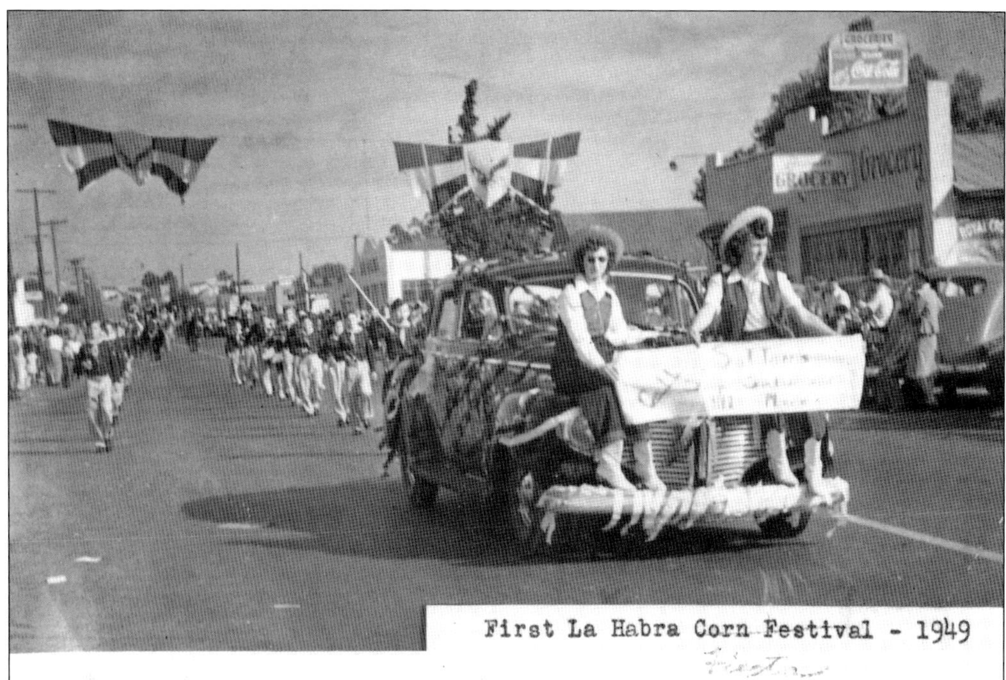

The photograph above captures the first-ever Corn Festival parade, which, along with the festival itself, began in 1949. The idea to build the beloved event around corn, never grown commercially in the city, is credited to La Habra Lion Bob Miller, who tapped into the nostalgia of summertime corn on the cob for the many Midwesterners who relocated to La Habra in the postwar era. The parade is considered one of the oldest and largest summer parades in Southern California, with an average of 25,000 attendees and 150 entries lining the mile-long route, such as this playful giant corn float entered by the Lions. Past parade grand marshals include National Football League champion Rosey Grier, outfielder on the 1981 World Series–winning Dodgers Jay Johnstone, and *Miami Vice* actor Edward James Olmos. (Both, courtesy of the La Habra Historical Society.)

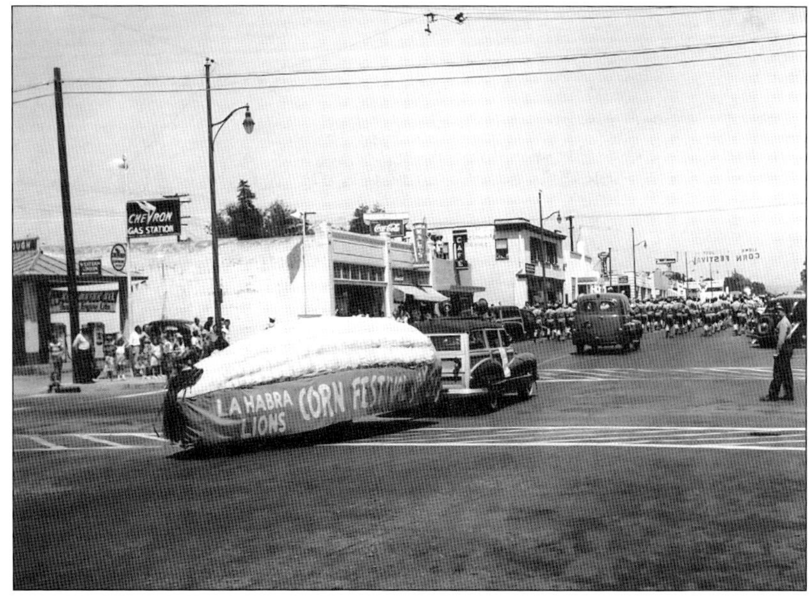

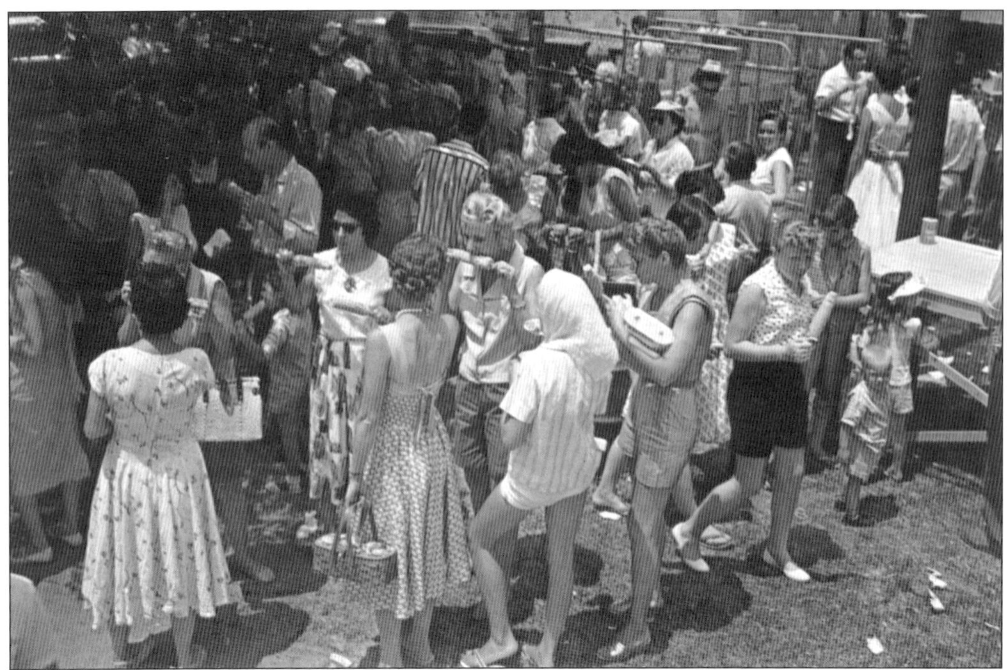

Corn Festival attendees take big bites out of the festival's signature cobs in this postcard from the 1950s. It is estimated that each year, the highly anticipated festival attracts about 75,000 people who consume around 14,000 ears of corn over the weekend-long event. (Courtesy of the Orange County Archives.)

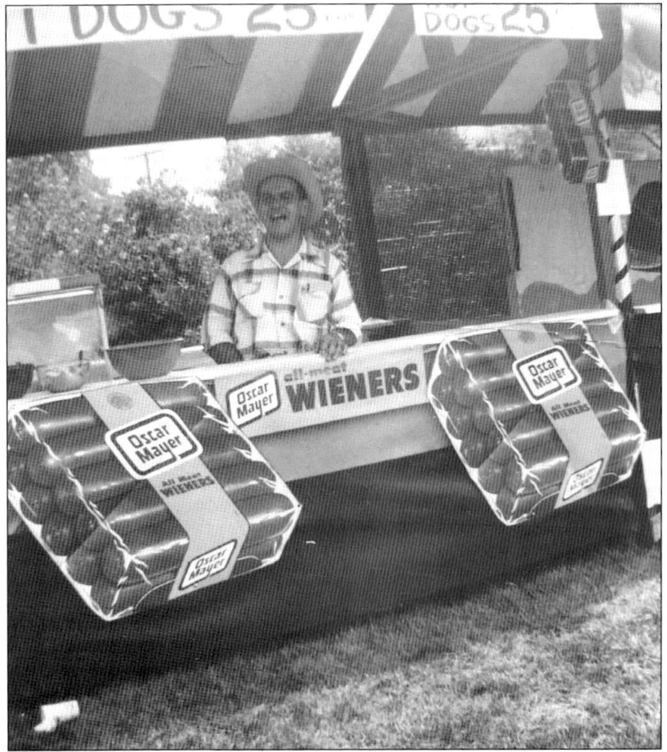

Tony Villelli sells hot dogs at a Corn Festival in the 1950s. After being released as a prisoner of war from a German war camp in June 1945, Villelli moved to La Habra and opened the La Habra Heights Café. In addition to becoming a realtor, Villelli served as a school board member, planning commissioner, Rotarian, Lion, and Our Lady of Guadalupe Knights of Columbus member. (Courtesy of David DeLeon.)

Along with the corn on the cob sold by the La Habra Host Lions, many of the food stands at the Corn Festival have been run by additional collaborating nonprofits, such as in this photograph of a root beer stand operated by the La Habra Woman's Club and its president, Mary Ellen Musson (right), in the mid-1990s. (Courtesy of the La Habra Woman's Club.)

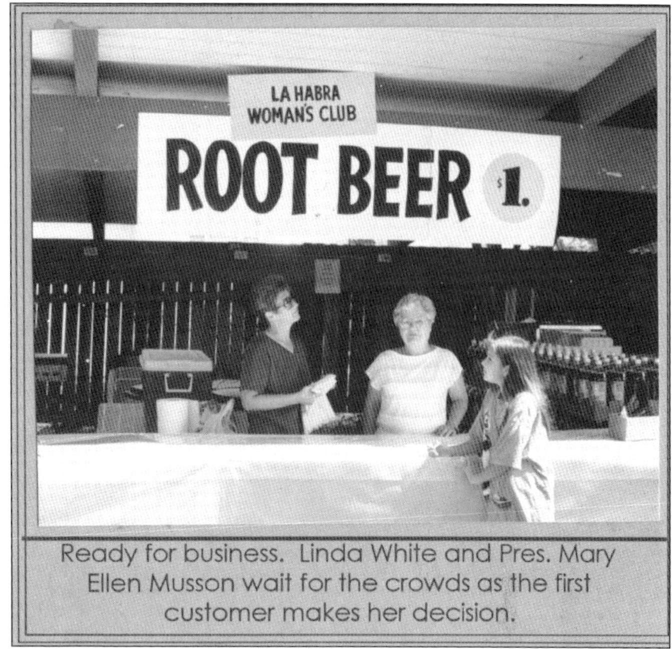

Ready for business. Linda White and Pres. Mary Ellen Musson wait for the crowds as the first customer makes her decision.

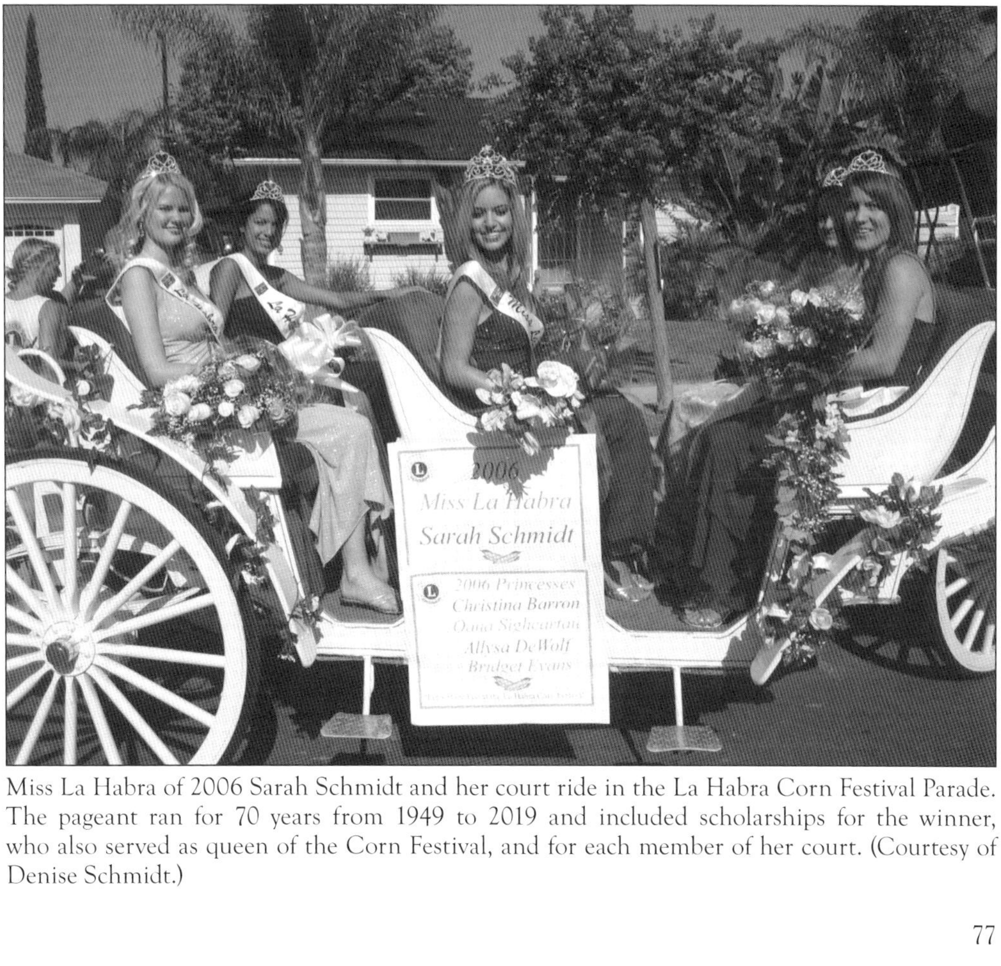

Miss La Habra of 2006 Sarah Schmidt and her court ride in the La Habra Corn Festival Parade. The pageant ran for 70 years from 1949 to 2019 and included scholarships for the winner, who also served as queen of the Corn Festival, and for each member of her court. (Courtesy of Denise Schmidt.)

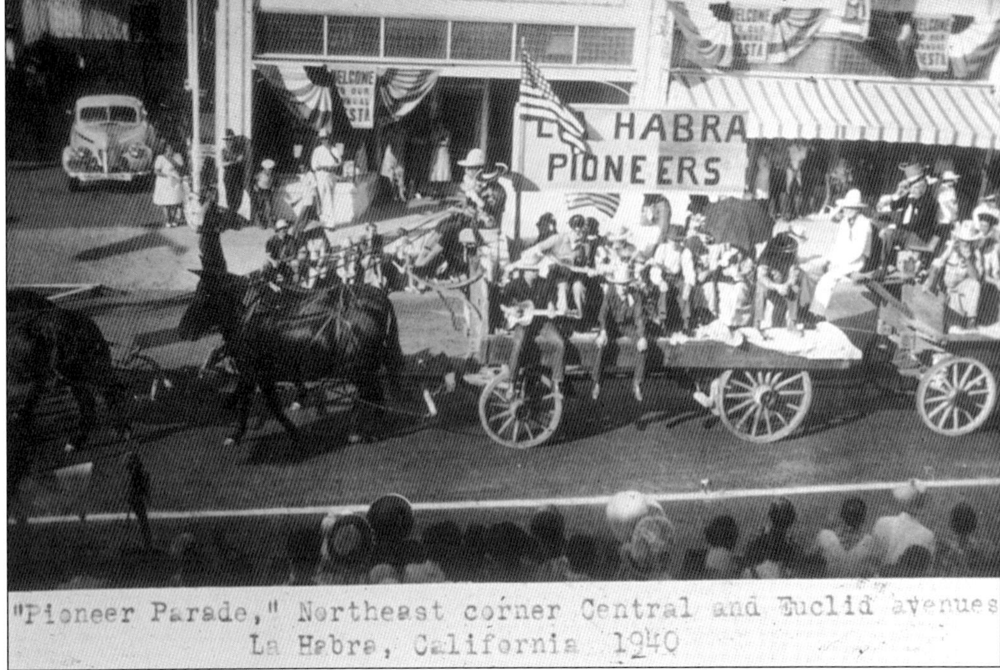

A float in the La Habra Fiesta parade was entered by the La Habra "Old Settlers," the precursor to today's La Habra Historical Society, in 1940. This photograph was taken on the northeast corner of Central and Euclid Avenue. The Fiesta parade was organized by the chamber of commerce and the La Habra Valley Riding Club, which met from the 1930s to the 1940s around Coyote Creek. (Courtesy of the La Habra Historical Society.)

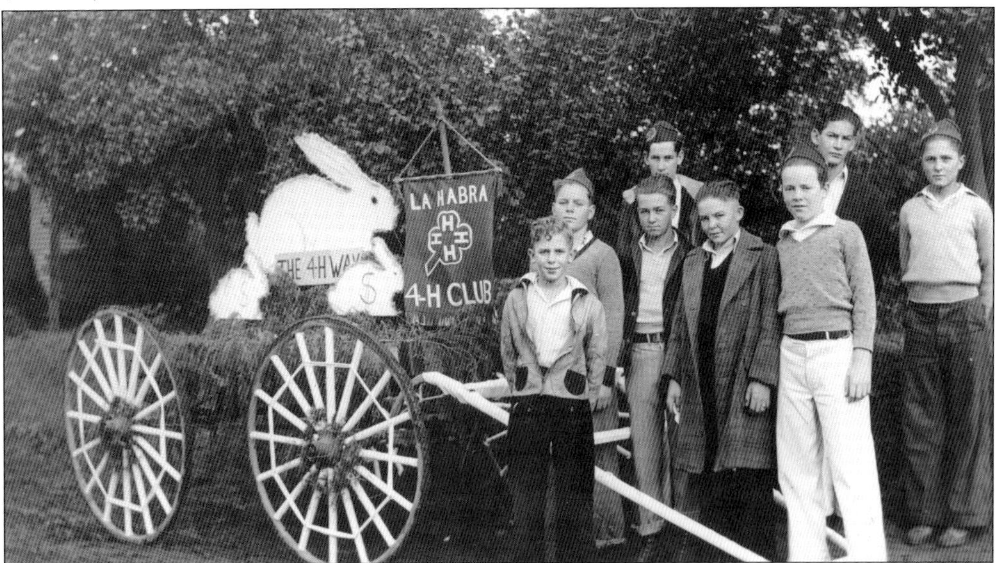

The 4-H Club stands next to its winning float from the 1934 La Habra Halloween Parade. Since 1972, the La Habra Hilltoppers 4-H Club has been part of the University of California Cooperative Extension Orange County 4-H Program and helps develop youth in La Habra Heights, La Habra, and surrounding communities. The four "Hs" stand for Head, Heart, Hands, and Health. (Courtesy of Orange County Archives.)

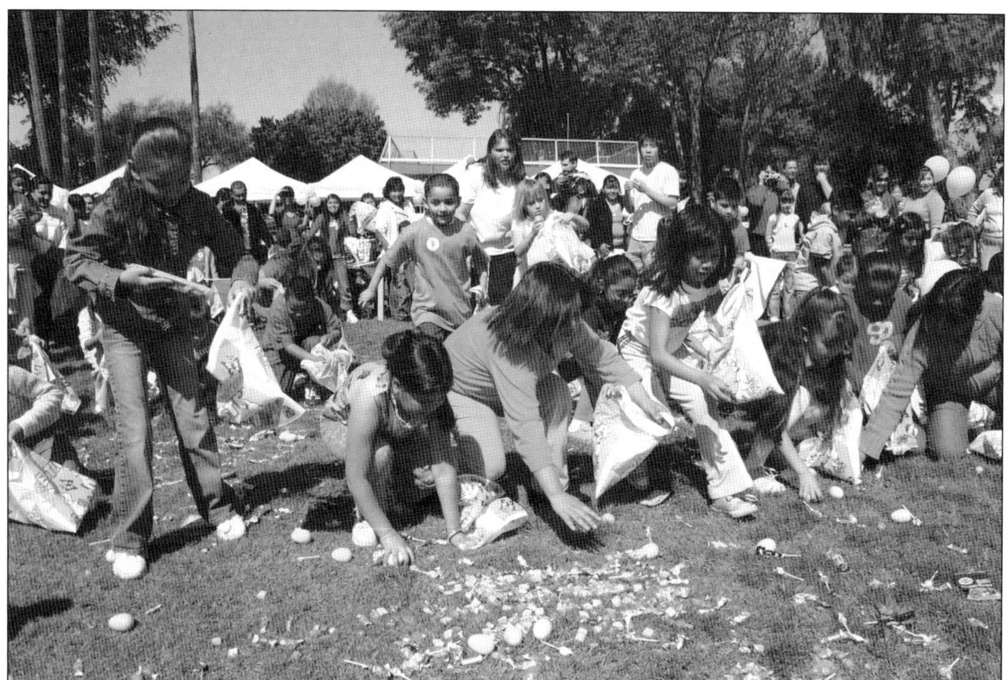

Kids scramble for eggs and candy in this photograph from the 2006 Spring Family Eggstravaganza. The event began in 2001 and includes game booths, the La Habra Lions' traditional pancake breakfast, and a free family health fair put on by the La Habra Family Resources Center, along with the egg hunts for various age groups. (Courtesy of the City of La Habra.)

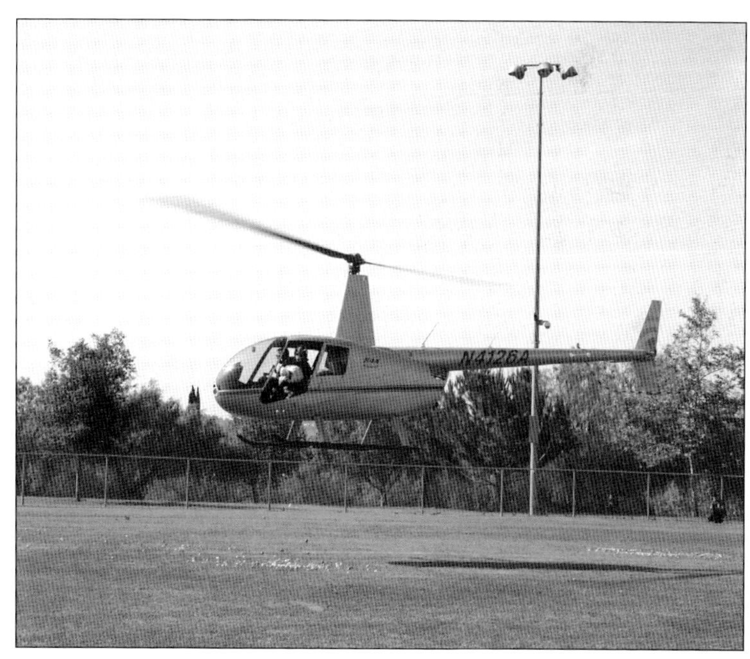

In 2014, the Spring Family Eggstravaganza added an exciting new feature to its festivities with the helicopter egg drop. Hundreds of colorful candy-filled eggs are dropped out of the helicopter window and onto the La Bonita Park outfield grass as excited kids wait for the copter to clear and for the hunt to begin. (Courtesy of the City of La Habra.)

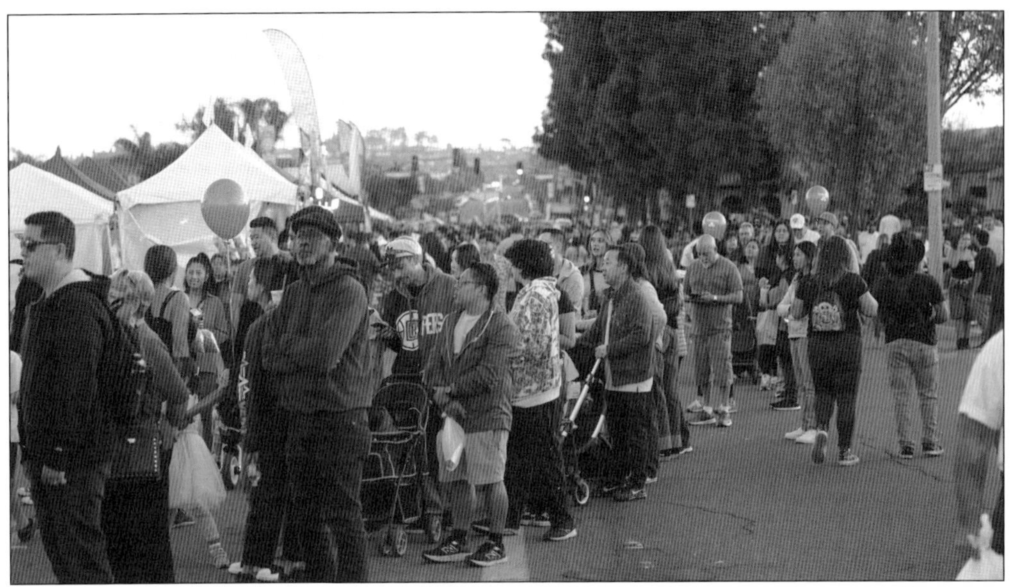

The Tamale Festival began in 2014 and quickly outgrew its first location in the parking lot of the former city hall. The all-day event now shuts down a section of Euclid Street to accommodate the nearly 10,000 attendees and over 20 tamale and food vendor stalls. (Courtesy of the City of La Habra.)

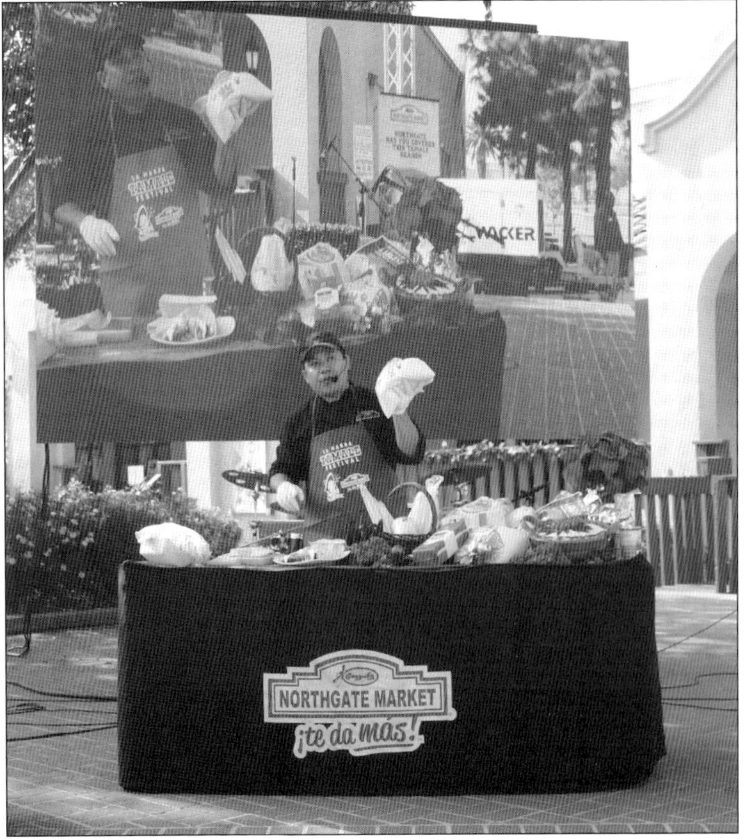

In addition to the tamale competition and food vendors, the Tamale Festival offers plenty of live entertainment, such as in this photograph of a tamale-making lesson from 2017 held on the main stage sponsored by Northgate Market in Portola Park. The event also boasts two additional stages—one at Brio Park and one at city hall. (Courtesy of the City of La Habra.)

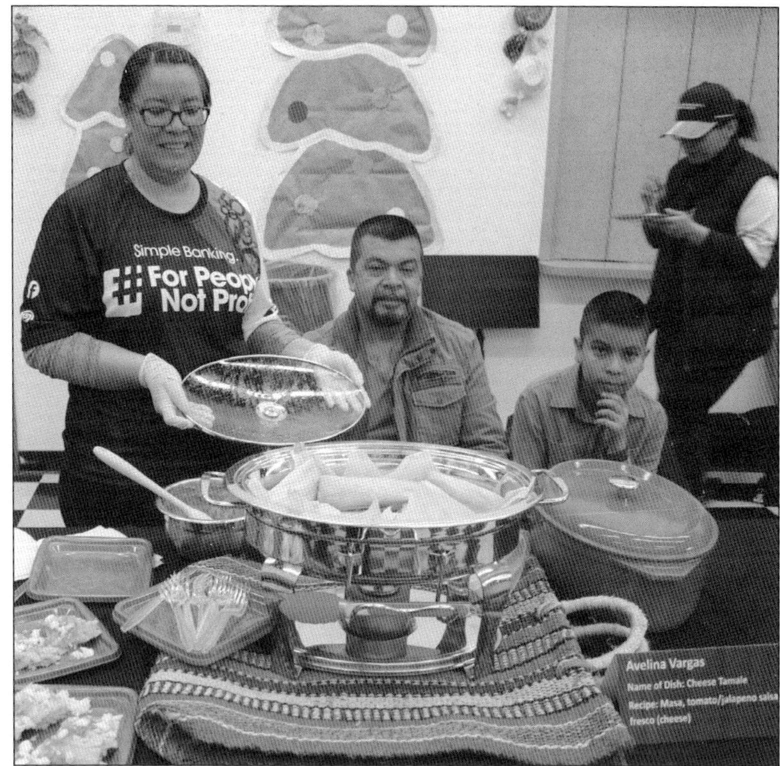

Tamale Festival contestant Avelina Vargas shows off her entry of cheese tamales in 2016. (Courtesy of the City of La Habra.)

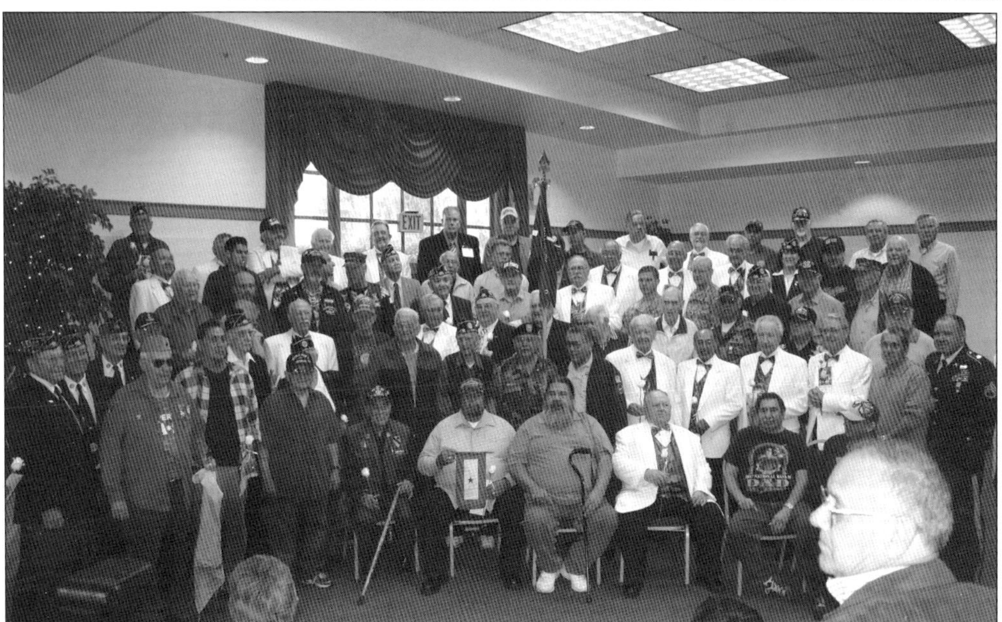

This group photograph was taken at the 11/11 Veterans Day of Remembrance event in 2011. For that year's commemoration, veterans were honored through song by the Orange Empire Chorus, discernible in this photograph with their bright white suit jackets, and fallen service members of Orange County were remembered through miniature memorials of white crosses and American flags that lined the room. (Courtesy of Bruce Martin.)

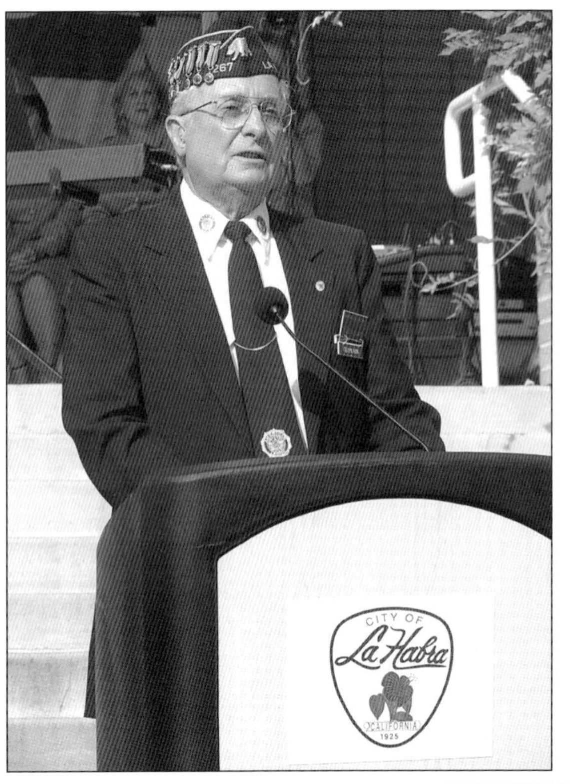

American Legion Post commander J.J. Dodge (below) and first vice commander Nile Stuart (left) reached out and partnered with the City of La Habra to create what are today's Veterans Day events, including the 11/11 at 11:00 a.m. Remembrance. These two veterans also helped bring Red Shirt Friday, in which red is worn on Fridays to show respect for the country's troops and veterans, to the city and worked to establish the Military Banner program in 2006 to honor active-duty residents. The banners, which can be seen on the light poles lining La Habra Boulevard, were originally displayed for one year from the date of installation. In 2014, the Veterans Committee expanded the banner program to include veterans, and each banner is now displayed for two years. (Both, courtesy of David DeLeon.)

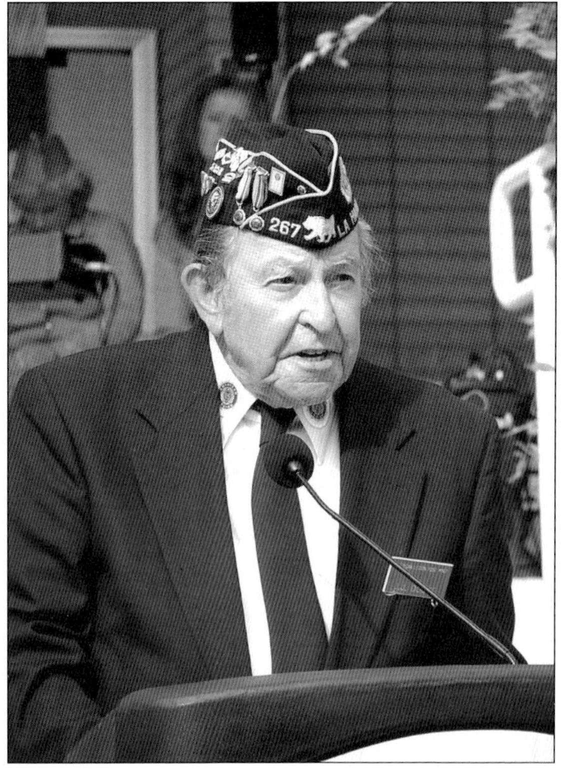

Eight

A Caring Community

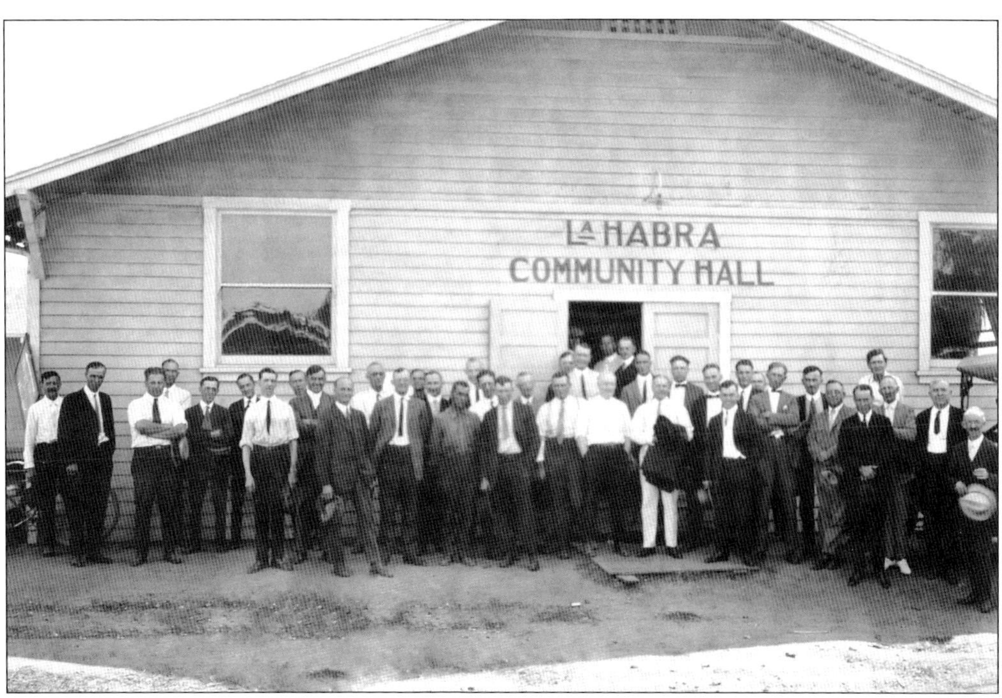

The admirable spirit of community service that has defined La Habra's Kiwanis Club for over 100 years can be observed in this photograph capturing charter members at a Kiwanis meeting in 1922. United by a shared commitment to bettering their community, the La Habra Kiwanis Club's current projects include promoting housing stability and food access, developing youth leaders, building playgrounds, and much more. (Courtesy of the City of La Habra.)

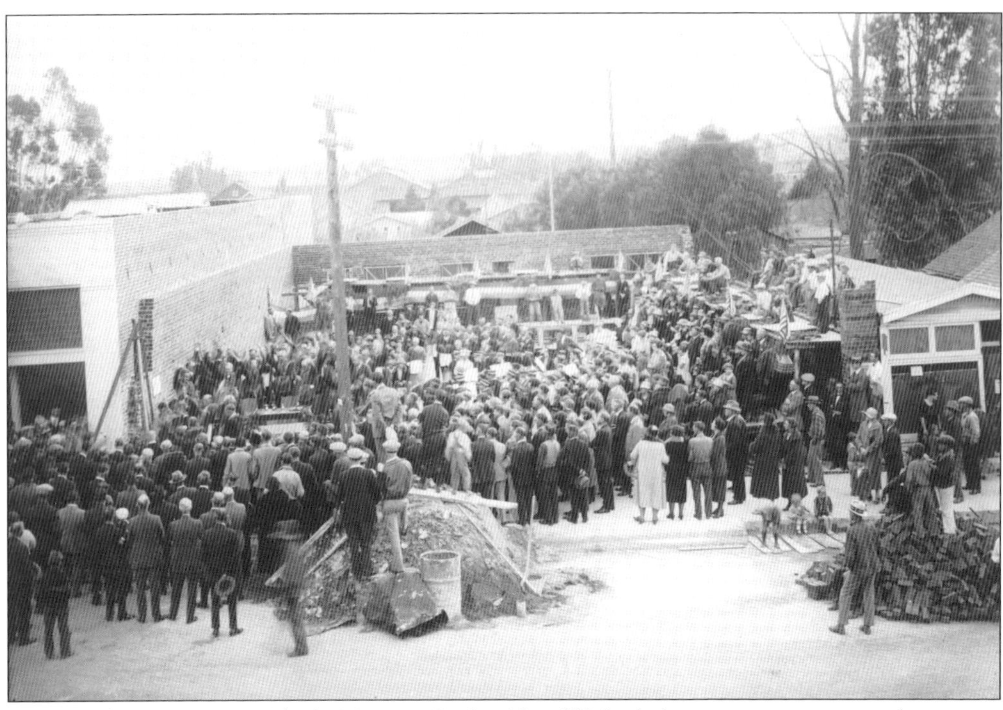

Chartered in 1927, La Habra's Masonic Lodge No. 659 laid the cornerstone on its large two-story building on the south side of the 100 block of Central Avenue in the late 1920s. An earlier attempt to organize a Masonic lodge occurred in 1923, but it was not successful until four years later. The group was famously a staunch supporter of public schools, presiding over cornerstone-laying ceremonies such as that for Lincoln School and for the Lowell District administration building that replaced it in 1968 and for years sponsored Public Schools Week. (Both, courtesy of the La Habra Historical Society.)

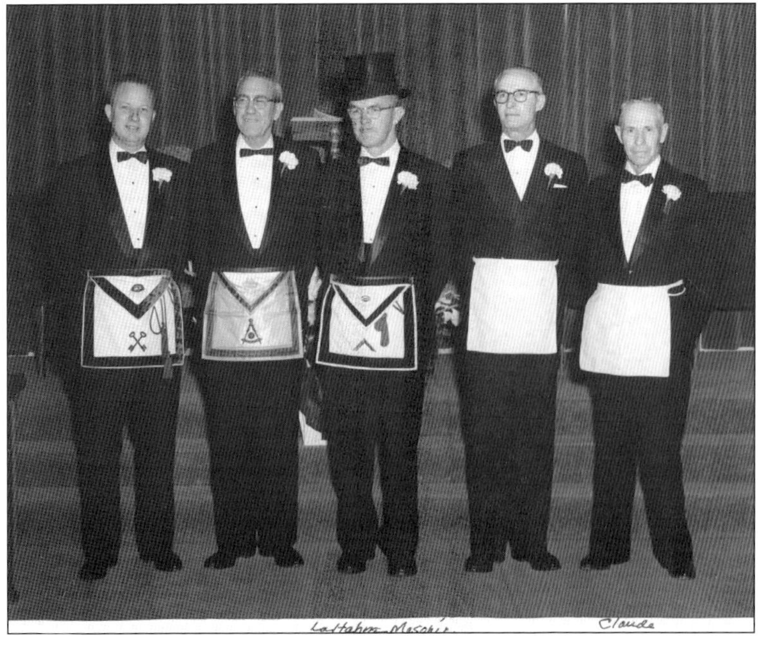

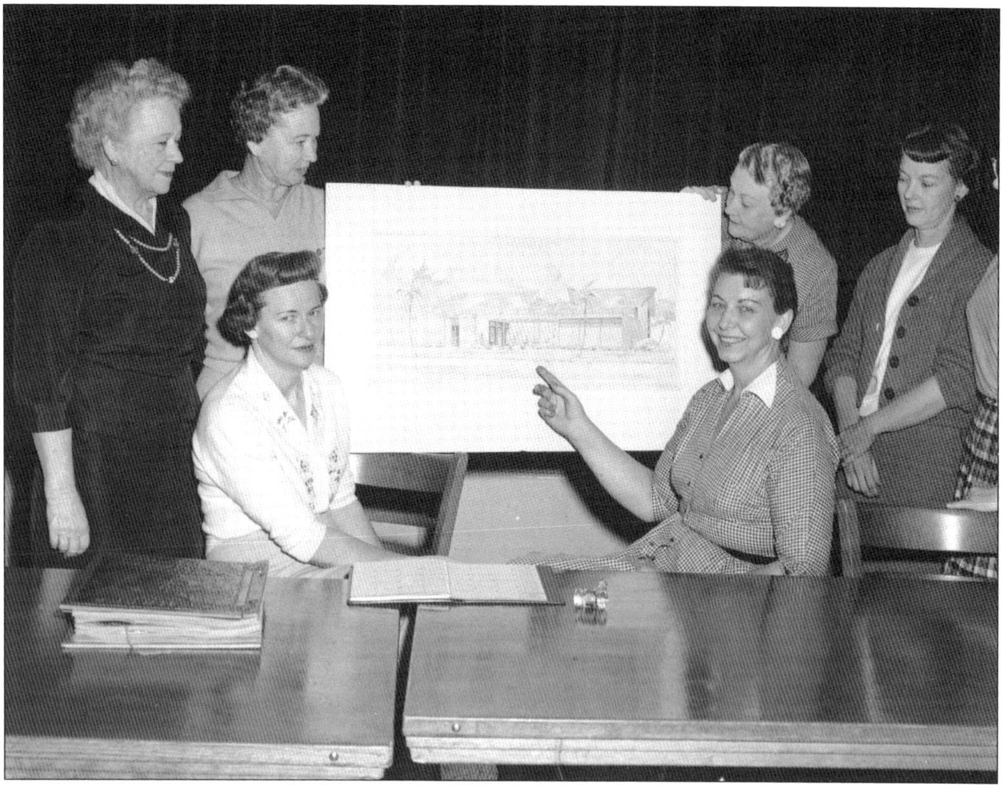

Members of the La Habra Woman's Club show off designs for a renovation project of their clubhouse, which began in 1961. The clubhouse, seen in the image below, was built in 1923 and was originally on Greenwood Avenue and Lois Street. The land was eventually donated to the city, and the city operated it as a community center for many years. A park named after the club and in honor of its dedicated service to the community will be opened in 2025. Today, the club meets in the annex of the city's current community center. (Both, courtesy of the La Habra Woman's Club.)

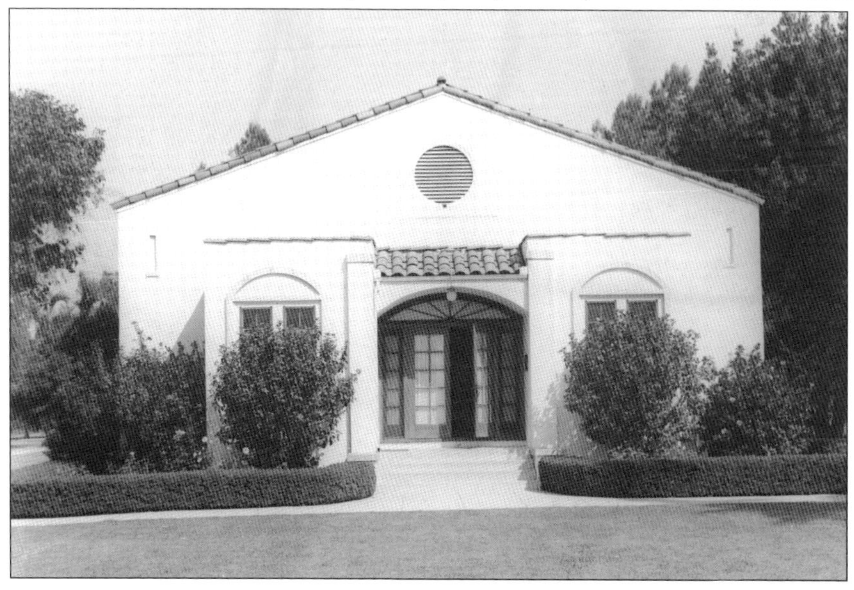

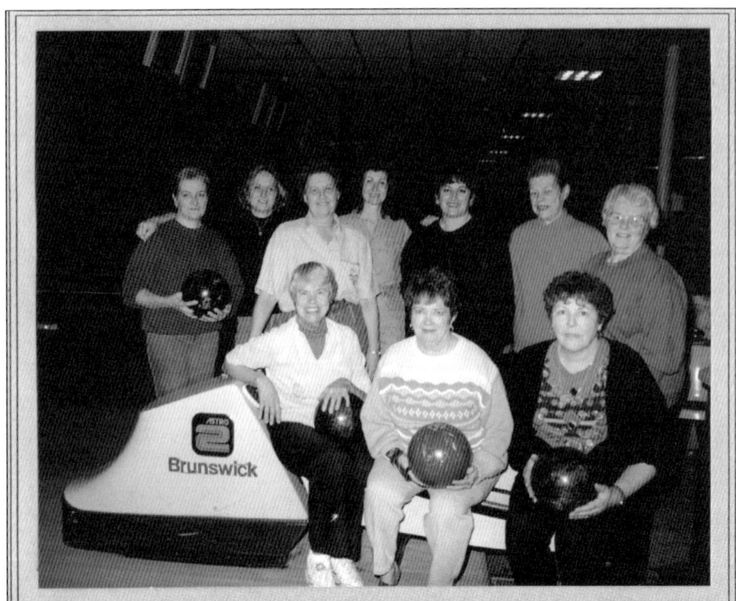

The La Habra Woman's Club bowling team earns funds for the Boys & Girls Club of La Habra in this 1994 photograph. The club has been influential in the city since its founding in 1898 and federation in 1912. It was instrumental in bringing the Boulevard of Bells to the city, naming many of the streets, and volunteering its time to charitable causes. (Courtesy of the La Habra Woman's Club.)

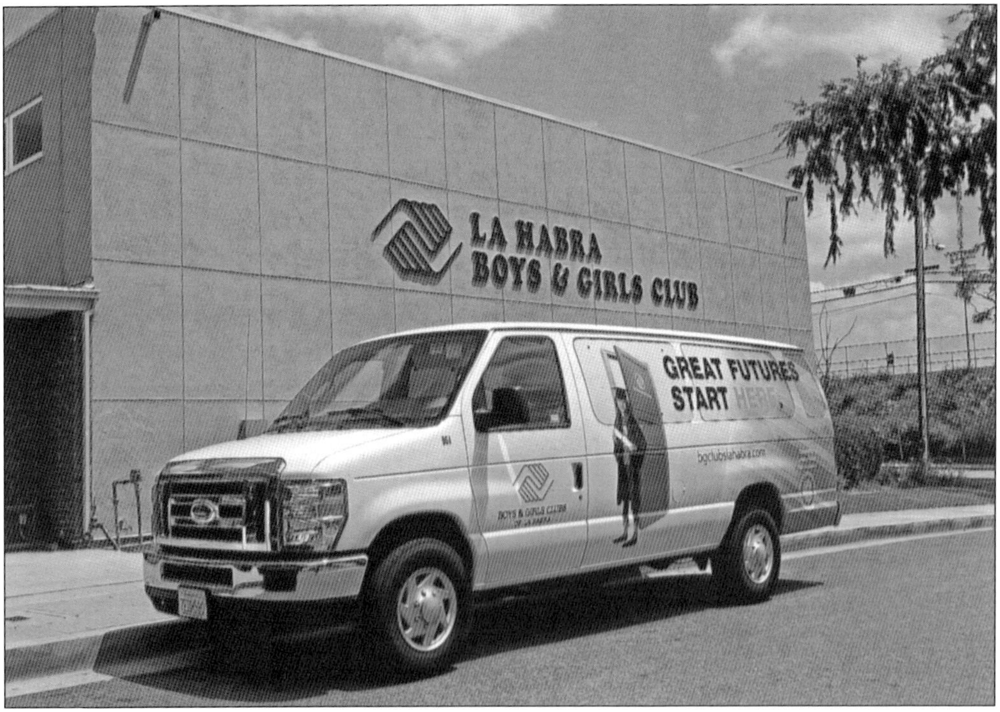

Founded by Tony Villelli, Al Brown, John Kelley, and the La Habra Host Lions Club, the Boys Club opened in 1958, and a wing designated as the Girls Club was added to the building in 1986. The mission of the club remains unchanged: help all young people, especially those who need it the most, reach their full potential as productive, caring, and responsible citizens. (Courtesy of the City of La Habra.)

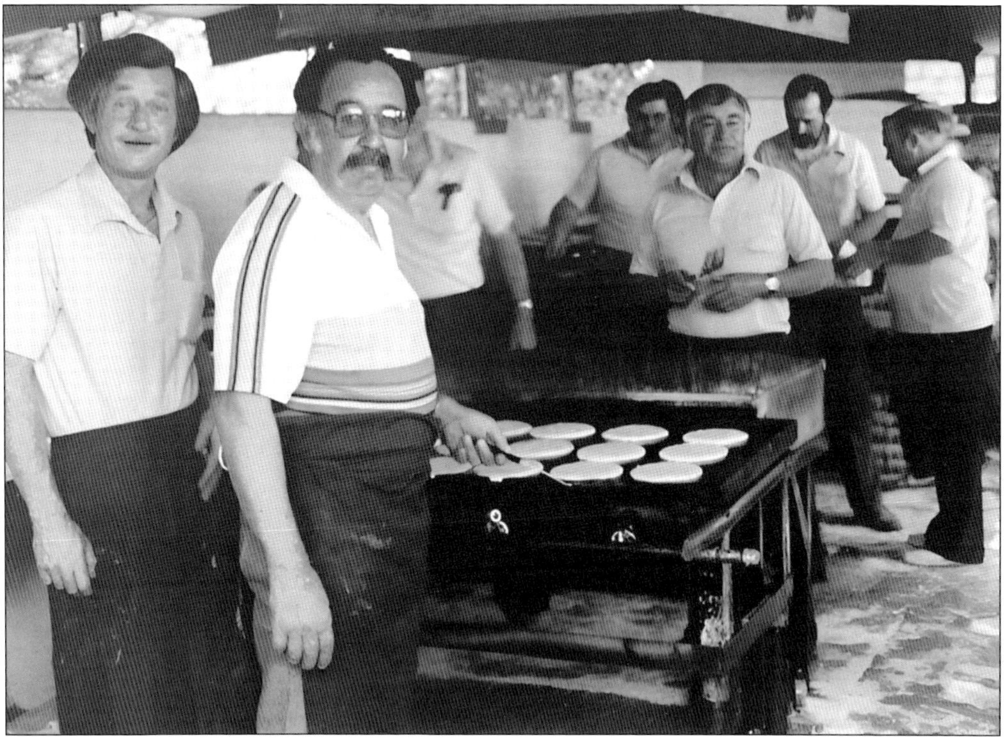

Pictured above, the La Habra Host Lions work the griddle at a pancake breakfast in the 1970s. Though some may only know the Lions from the annual Corn Festival and their funnel cake booth seen at many city events, the club has been a cornerstone of La Habra's social and philanthropic landscape since its founding in 1947. The Lions volunteer for many organizations and causes, but the two main focuses for their fundraising work in recent years are youth and school programs and public health initiatives. Pictured below, the Lions cook lunch for Love La Habra volunteers in 2015. (Above, courtesy of David DeLeon; below, courtesy of the City of La Habra.)

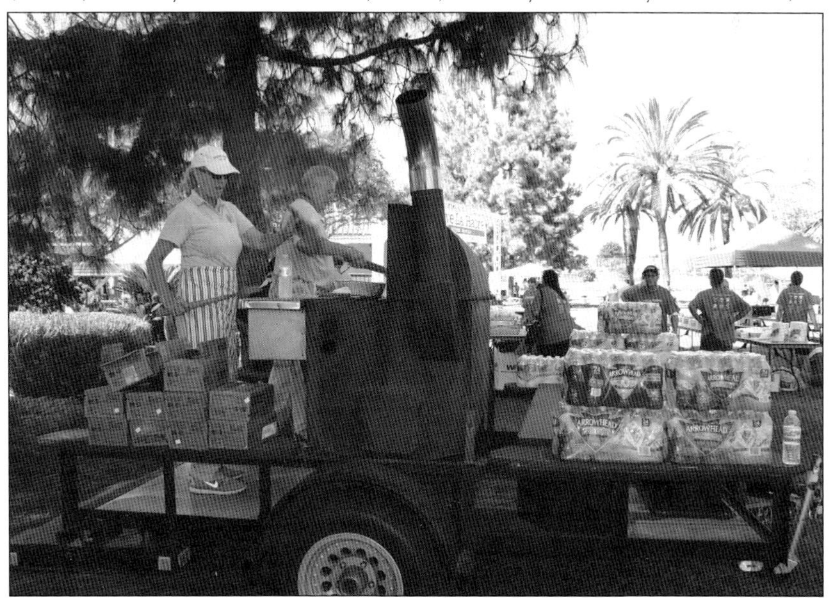

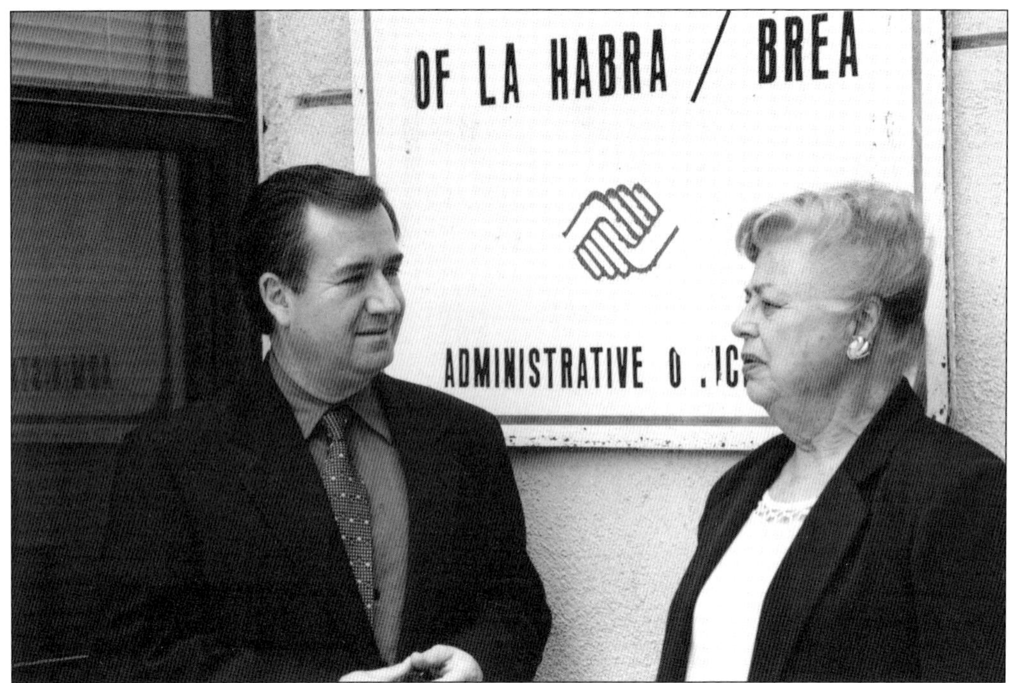

Esther Cramer (right) is known as the most knowledgeable historian to document the growth of La Habra, writing *La Habra: The Pass through the Hills*, a seminal history of the town's early days. She was appointed La Habra's 50th-birthday celebration chairwoman in 1975 and was also active with the Boys & Girls Club and the Children's Museum. (Courtesy of the Boys & Girls Club of La Habra.)

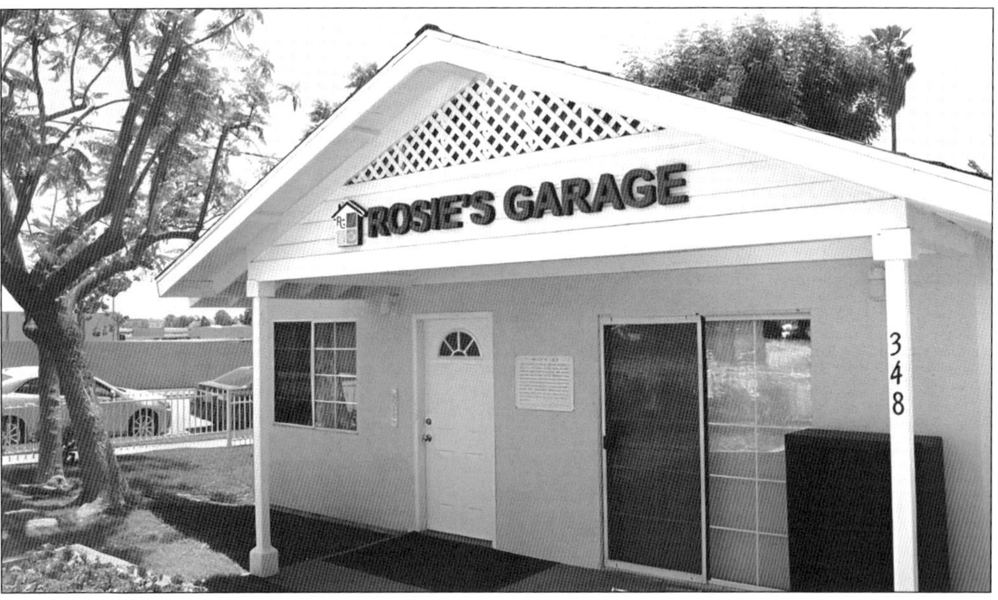

In the late 1980s, the city experienced a surge in gang-related violence that disrupted the lives of children, particularly those left unsupervised after school. In 1991, Rose "Rosie" Espinoza opened her doors to children within her community as a safe space to do homework while they waited for their parents. By 1996, the organization had seen a reduction in local crime and an increase in graduation rates. (Courtesy of Rose Espinoza.)

Inspired by a friend who struggled to care for a child with special needs, Cleta Harder founded Help for Brain Injured Children Inc. (HBIC) in 1967. In 1980, HBIC established the Cleta Harder Developmental School in a small home; a larger facility was built in 2023. The nonprofit organization has maintained its mission of "Giving Hope" to individuals with disabilities and support to their families. (Courtesy of HBIC.)

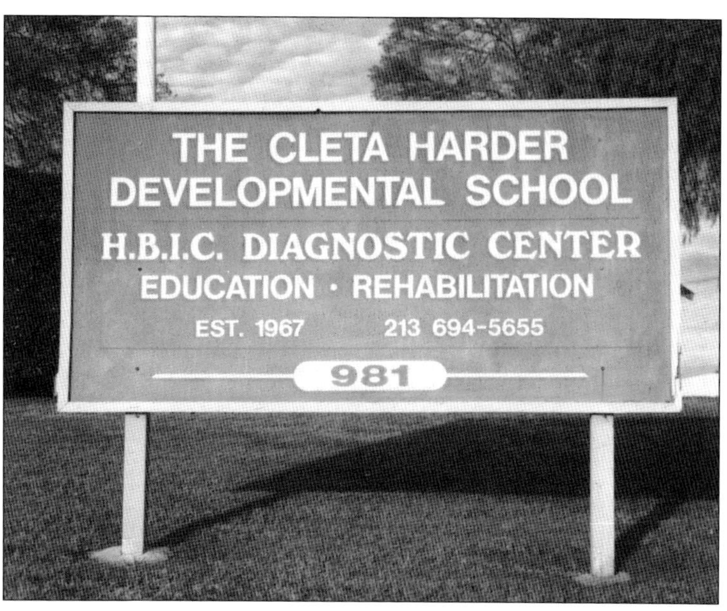

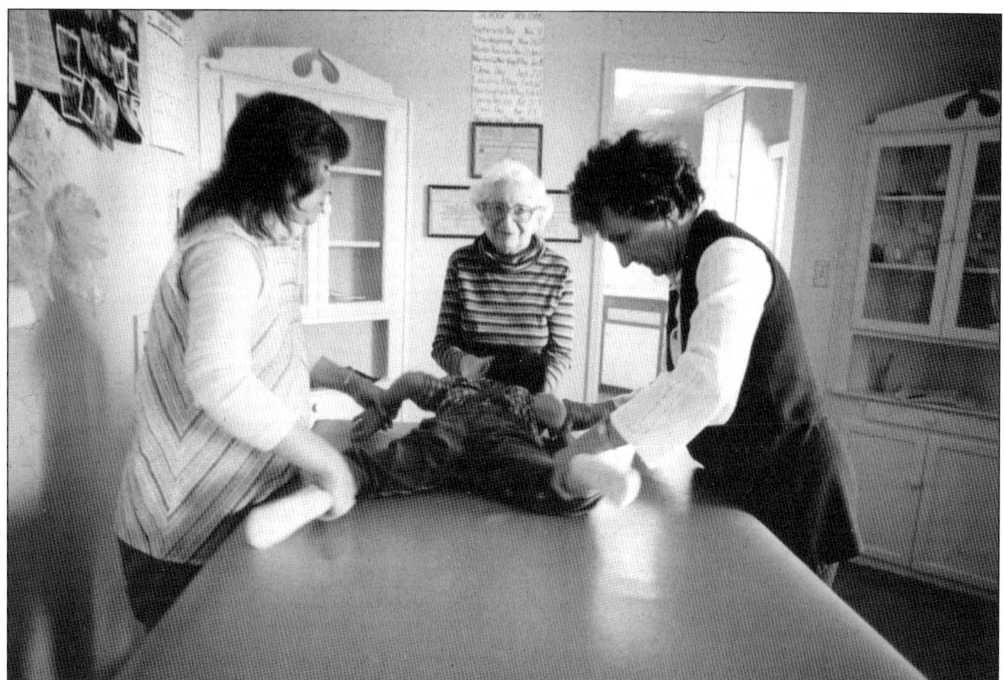

Patterning was a form of therapy used for treating the children participating in the HBIC alternative education program. Volunteers like these were the primary employees during HBIC's early years, allowing community members, students, and parents hands-on experience working with the special needs population. Opportunities for these groups still exist today, though in a more academic setting. (Courtesy of HBIC.)

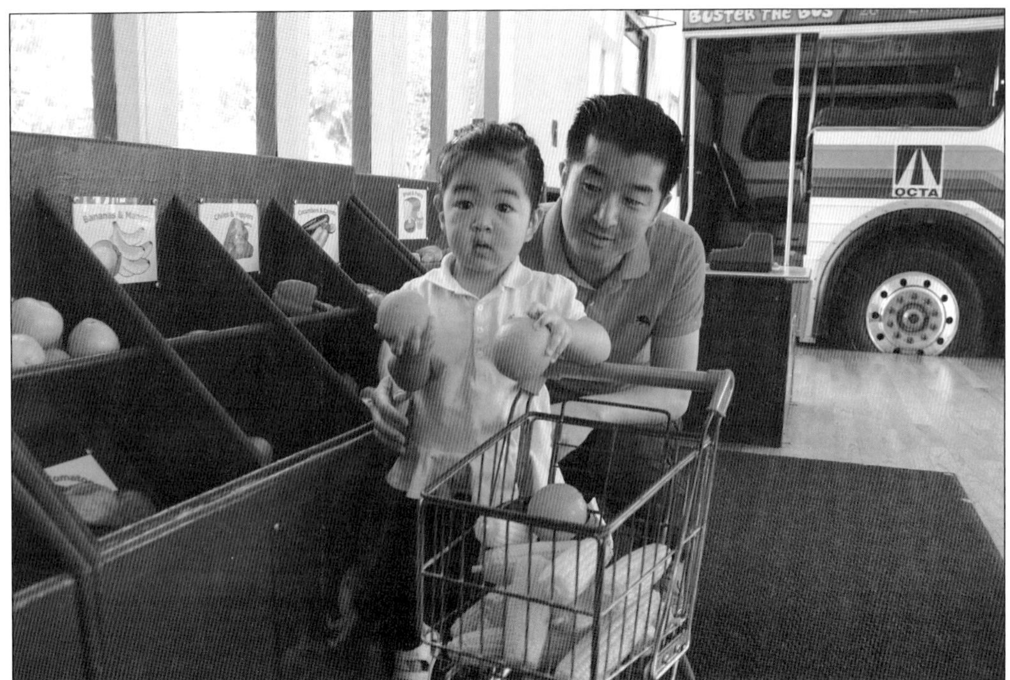

In the photograph above from 2016, a father and daughter play in the Mini Market, one of many exhibits offered by the Children's Museum at La Habra. Housed in the former Union Pacific train depot, which ceased operations in 1950, the museum became one of the first of its kind on the West Coast when the converted depot's doors were reopened in 1977. Today, the museum is visited annually by 95,000 children who get to experience seven interactive and educational galleries, including the Kids on Stage exhibit, as seen in the photograph below. (Both, courtesy of the City of La Habra.)

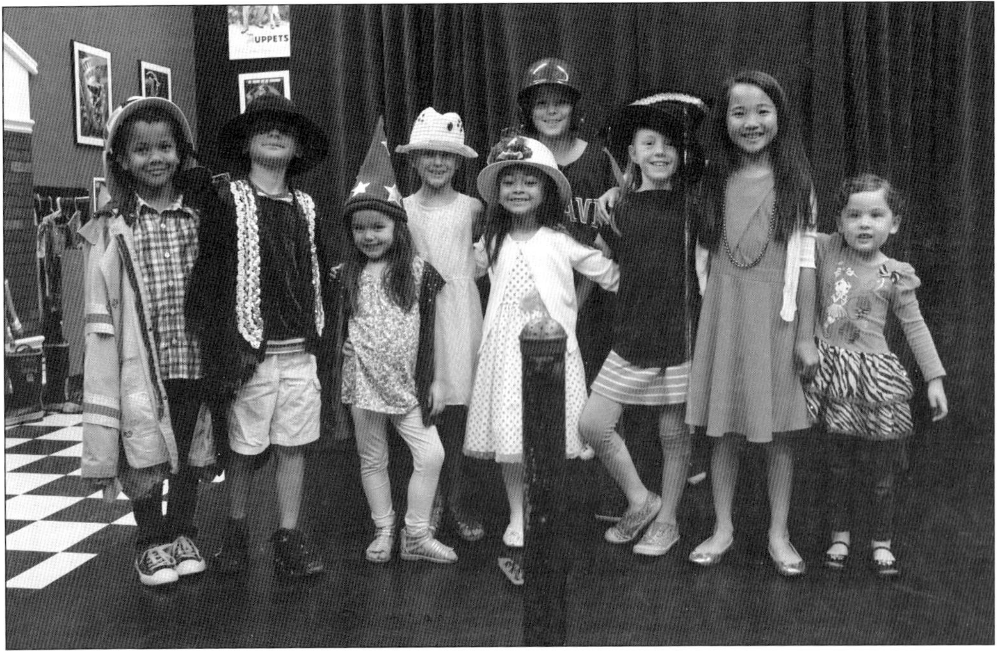

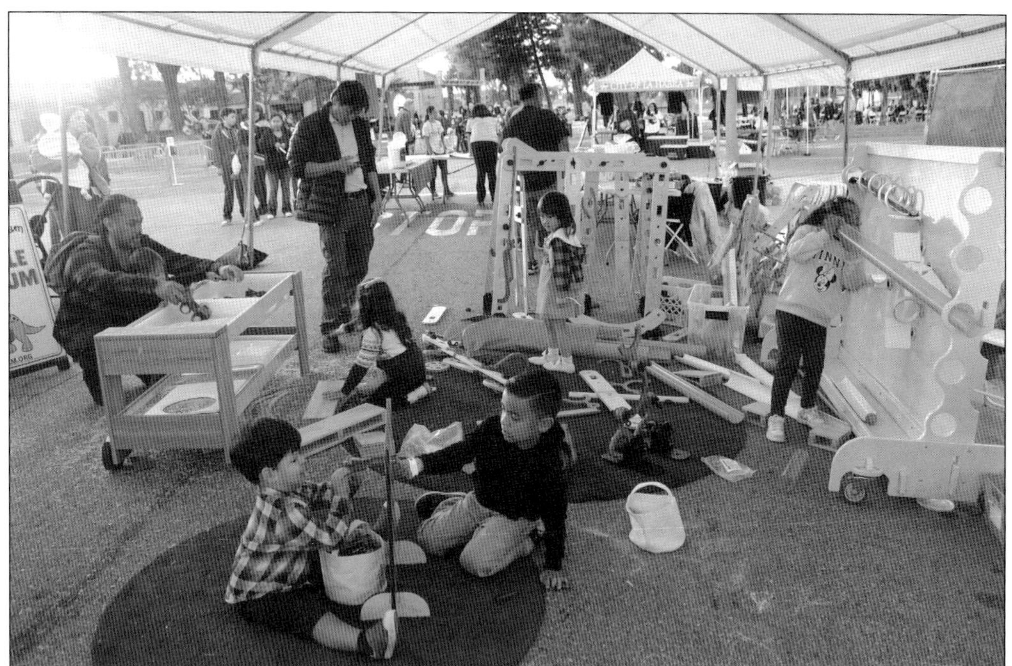
An extension of the Children's Museum, the Mobile Museum often travels to city events to bring a playful education to children of the community, as seen at an unidentified city event pictured here. The museum on the go offers hands-on sensory play, trained museum facilitators, games and crafts, insect meet & greet, and more. (Courtesy of the City of La Habra.)

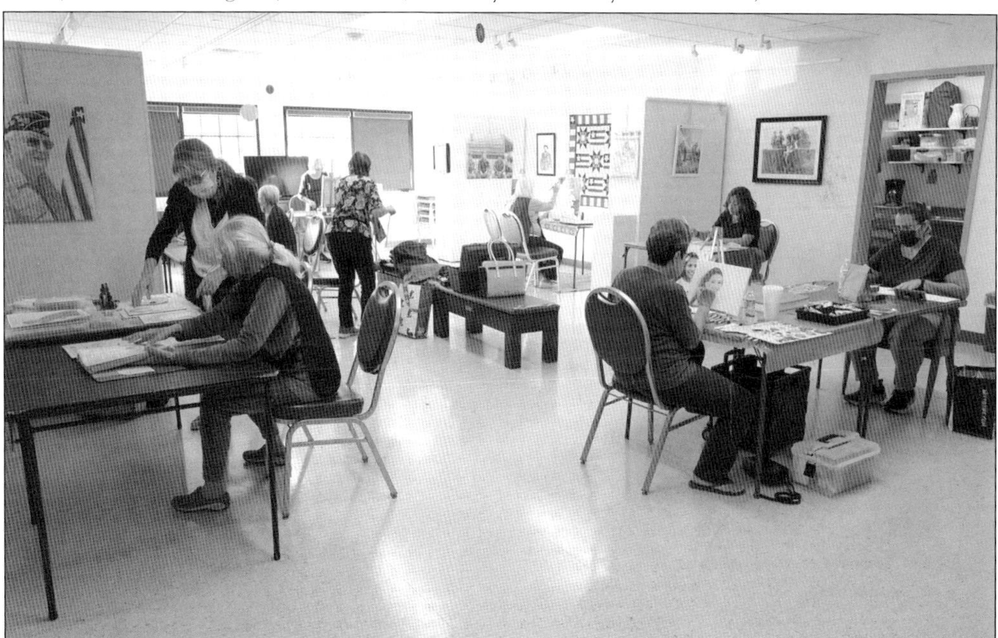
Founded in 1955, the La Habra Art Association provides artists with venues to showcase their work while promoting art appreciation in the community. In 2017, the La Habra Art Association opened the La Habra Art Gallery in the La Habra Community Center annex. Here, patrons are led in a painting class by a local art instructor. (Courtesy of the La Habra Art Association.)

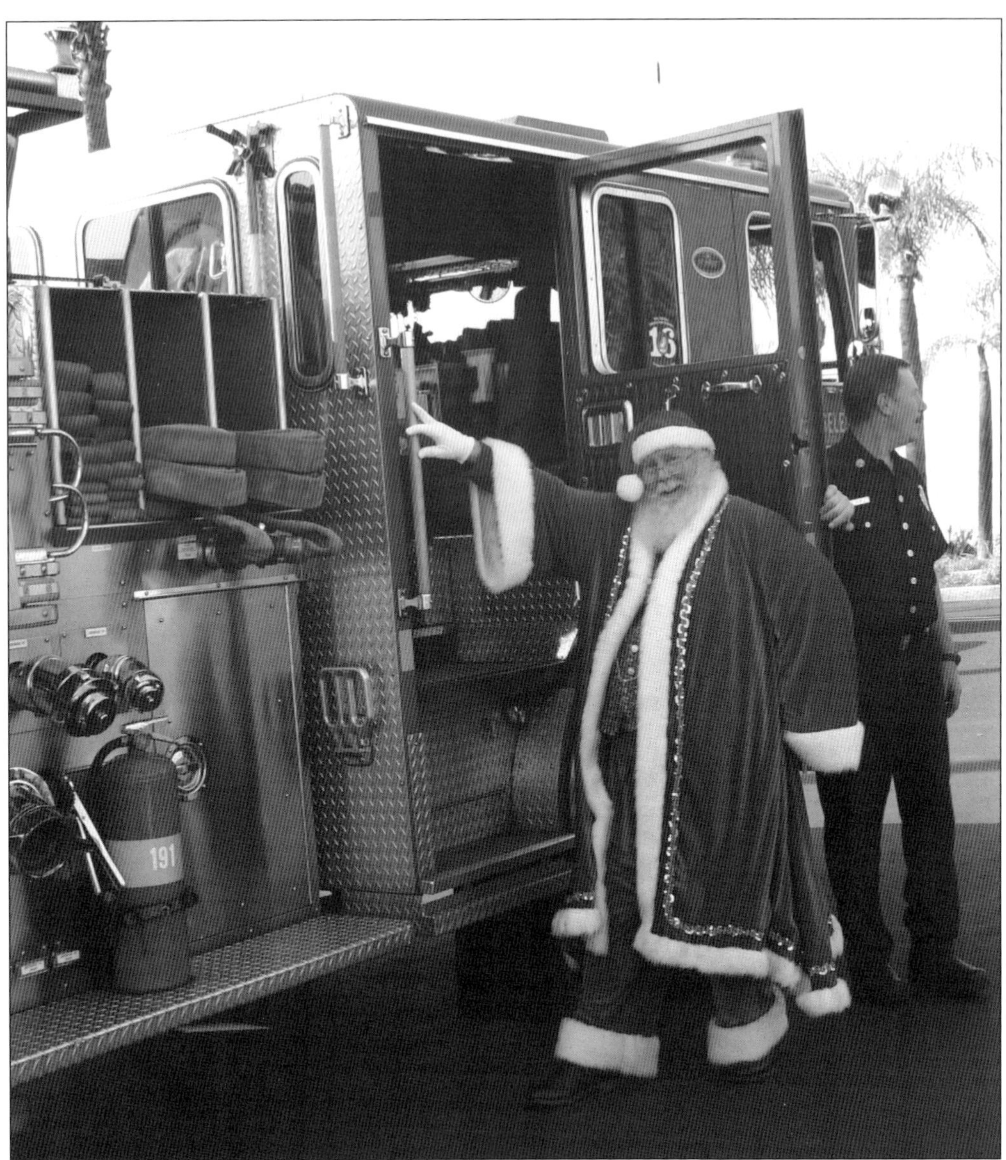

Every year, the City of La Habra Recreation Division and the community unite for the Holiday Wishes Program, formally known as Operation Santa, which has supported children and families facing challenging circumstances during the holiday season. In years past, the heartwarming community program culminated in an event where families visited with Santa and played carnival games, as seen here from Operation Santa in 2009. The La Habra Girls Softball Association has also hosted the Operation Santa Tournament, where each player participating is asked to bring a gift to donate toward the program. This collaborative effort demonstrates La Habra's commitment to supporting its residents and to making the holidays a little brighter for its residents in need. (Courtesy of the City of La Habra.)

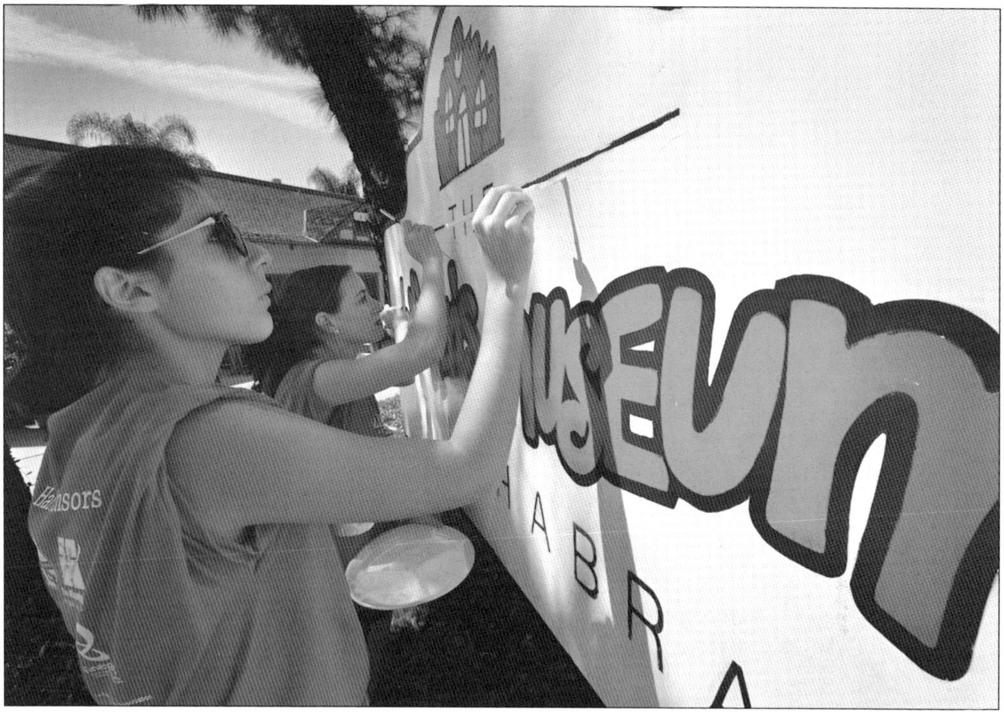

The city held the first Love La Habra Day of Service on September 26, 2015, joining the Love Our Cities movement that began in Modesto, California. Love Our Cities prioritizes practical ways to improve a city and its community. Each year, Love La Habra volunteers meet at Portola Park before splitting off to different sites across the city, such as in the photograph above, where a volunteer repaints the Children's Museum sign, restoring its bright colors, and in the photograph below of volunteers staffing the La Habra Community Resources Care Center. It is estimated that the work completed on these days is worth at least $18,000 in man-hours. Community members can submit ideas for projects ahead of the event, truly making this a celebration of collaboration in the caring community of La Habra. (Both, courtesy of the City of La Habra.)

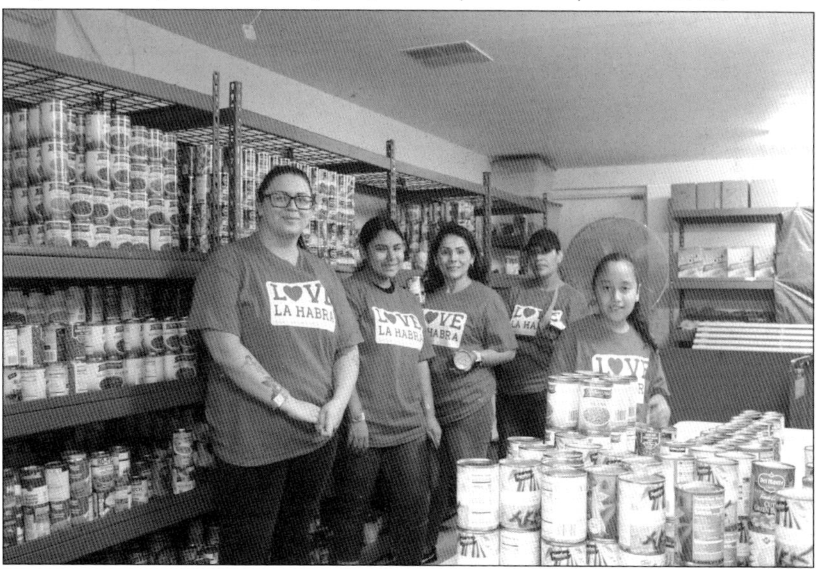

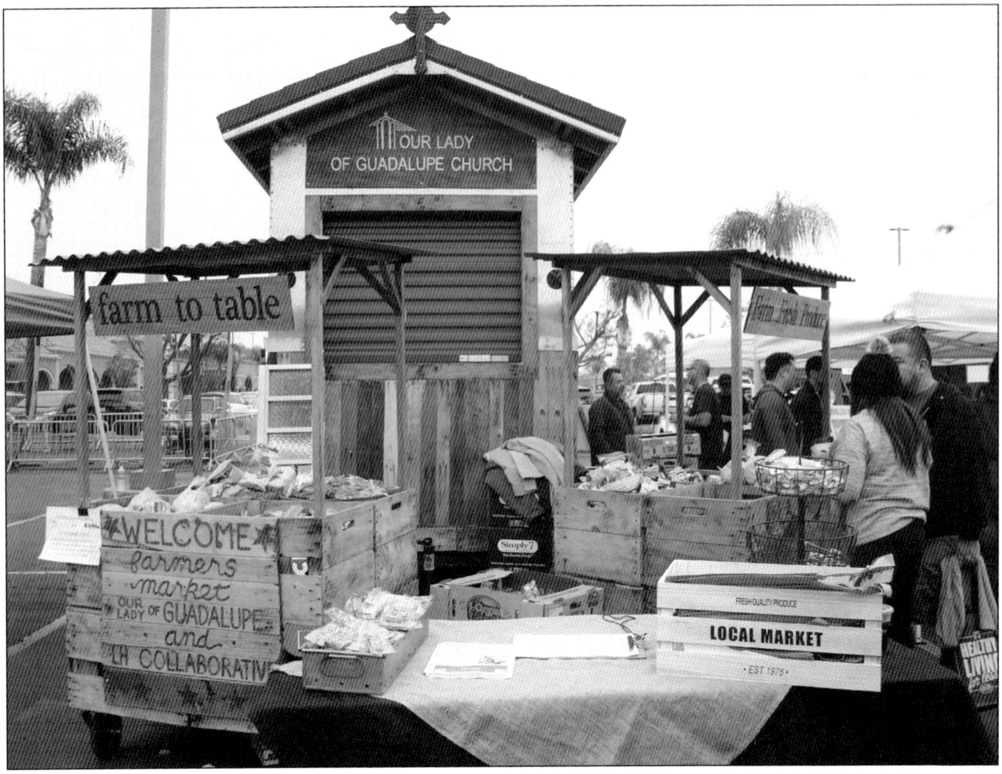

In recent years, Our Lady of Guadalupe has partnered with the La Habra Collaborative and Community Action Partners to host nutrition education classes and provide pop-up farmers' markets throughout the year, such as the one seen here in 2018. Sometimes free to attendees, the mission of these markets was to provide community members access to healthy foods. (Courtesy of the City of La Habra.)

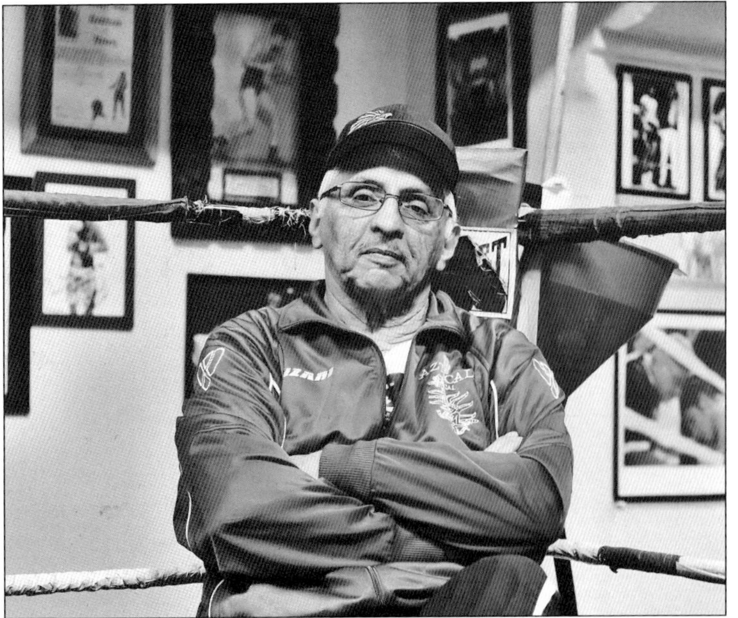

The La Habra Boxing Club, founded by David Martinez, provided training for thousands of youth for nearly 40 years before closing in 2020. The club was housed at the old Our Lady of Guadalupe Church and, for a time, was funded by the Boys & Girls Club of La Habra and the city. In 2009, Martinez started the La Habra Boxing Club Foundation, whose donations funded the operation. (Courtesy of David DeLeon.)

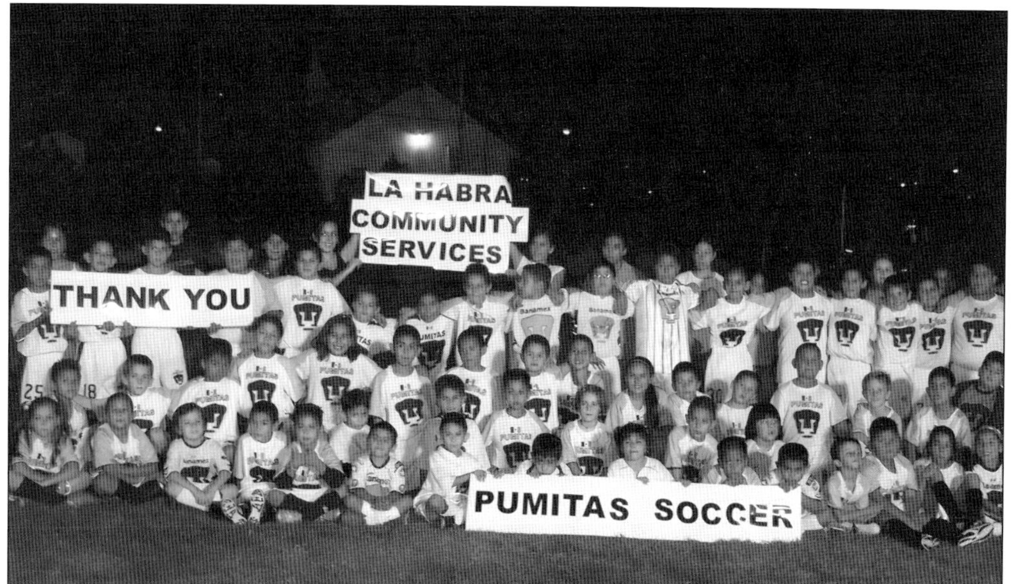

In 2006, youth soccer teams worked with the city to establish Pumas La Habra Soccer Club, the first club-level soccer program in the city. At the end of the first soccer practice at Portola Park, the club took a photograph to thank La Habra Community Services. As an affiliate of the City of La Habra, the club has participated in many local celebrations. (Courtesy of Pumas La Habra Soccer Club.)

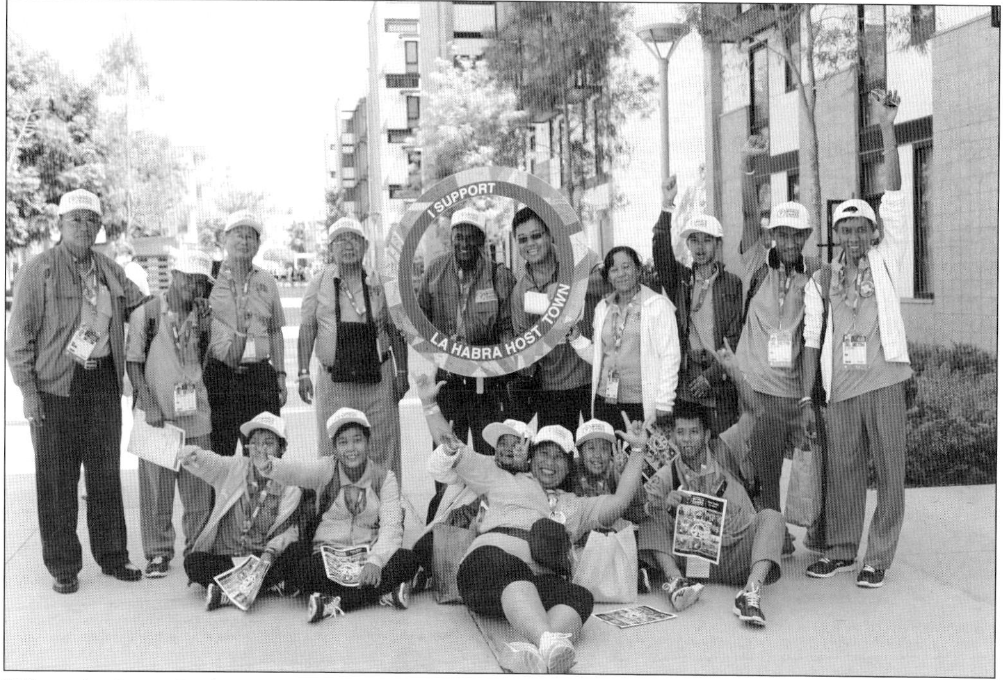

When the Special Olympics World Games were held in Los Angeles in 2015, La Habra was chosen to be a host town for the delegations from Lithuania, Norway, and Myanmar. The host town program is a three-day event prior to the opening ceremony and gives the global athletic delegations an opportunity to take part in a cultural exchange of activities. (Courtesy of the City of La Habra.)

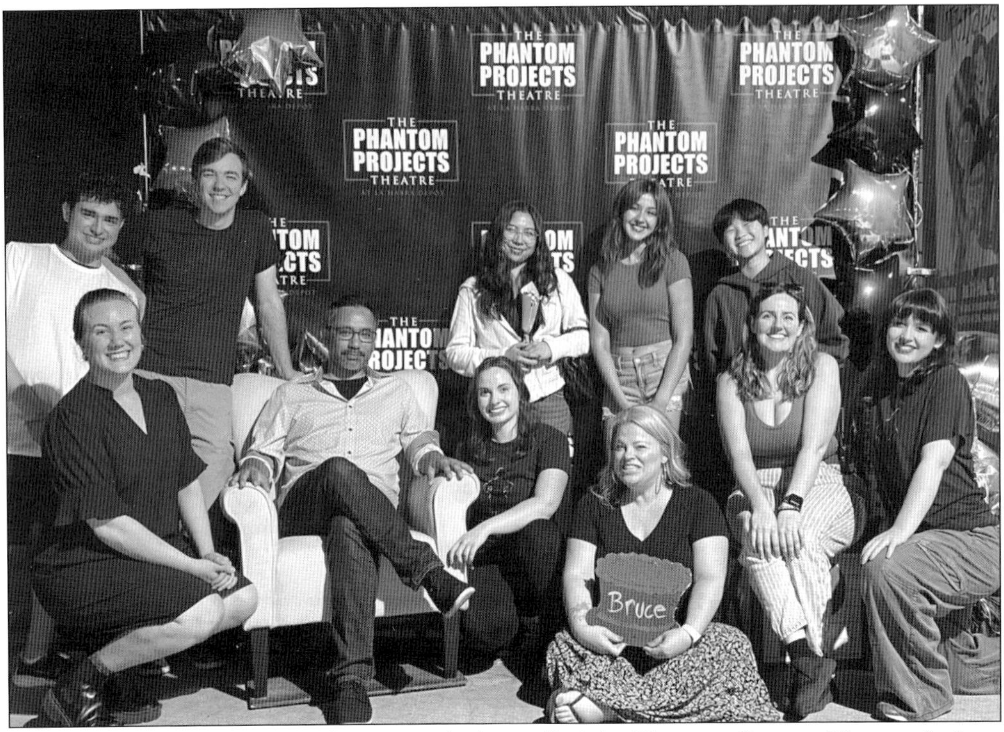

Founder Steve Cisneros (seated) poses with the staff of the Phantom Projects Theatre. In June 2021, Phantom Projects Theatre Group became the new manager, operator, and programmer of the historic La Habra Depot Theatre. In working with the city, the venue got a complete renovation; new carpet, flooring, paint, equipment, and more helped ensure that when the depot reopened in 2022, it was a high-quality venue in which to watch an array of performances. Below, students attend a field trip performance of *You're a Good Man, Charlie Brown* at Phantom Projects Theatre at La Habra Depot. Each year, thousands of local children attend these accessible and affordable shows that include sensory-friendly options. The theater works closely with the City of La Habra, school districts, nonprofit organizations, and businesses through community outreach and engagement. (Both, courtesy of Phantom Projects.)

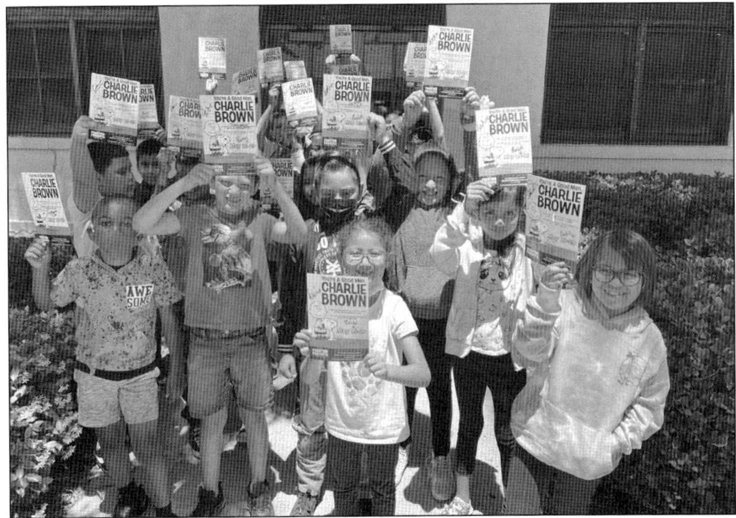

Nine

The Center of Civic Life

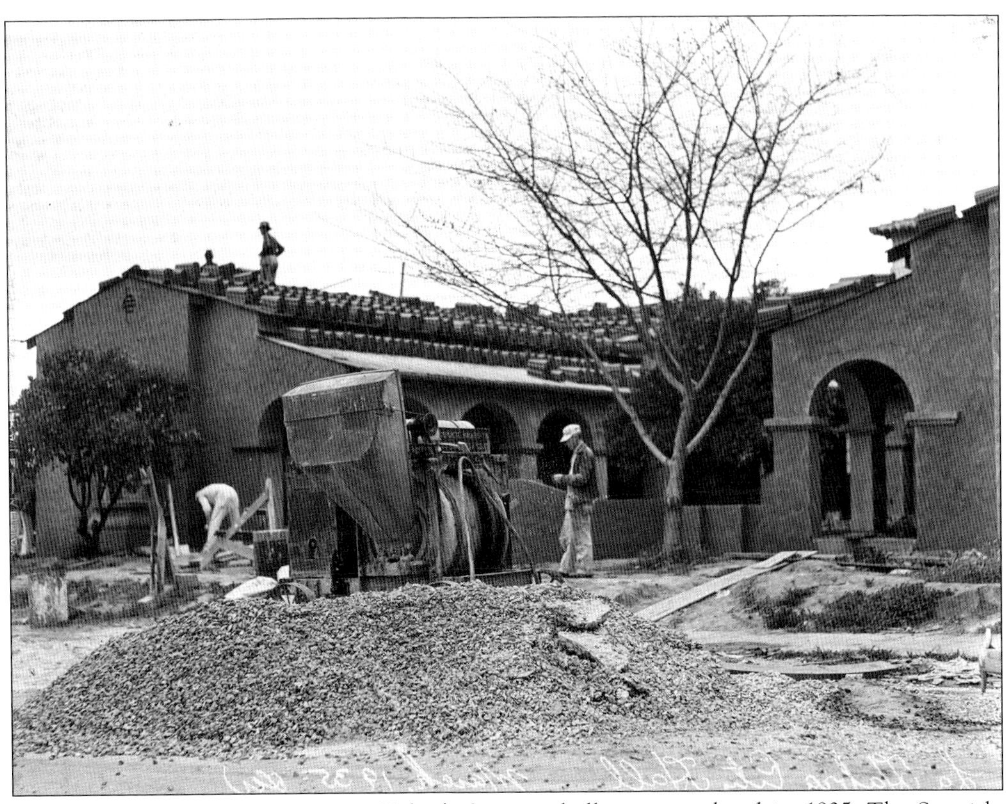

Seen here under construction, La Habra's first city hall was completed in 1935. The Spanish-style building was constructed of two wings flanking a courtyard. Prior to its construction, the city government had worked out of rented offices. In addition to city administrative offices, the civic center contained the library, the American Legion Hall, and the fire department before new stations were built in the 1960s. (Courtesy of the City of La Habra.)

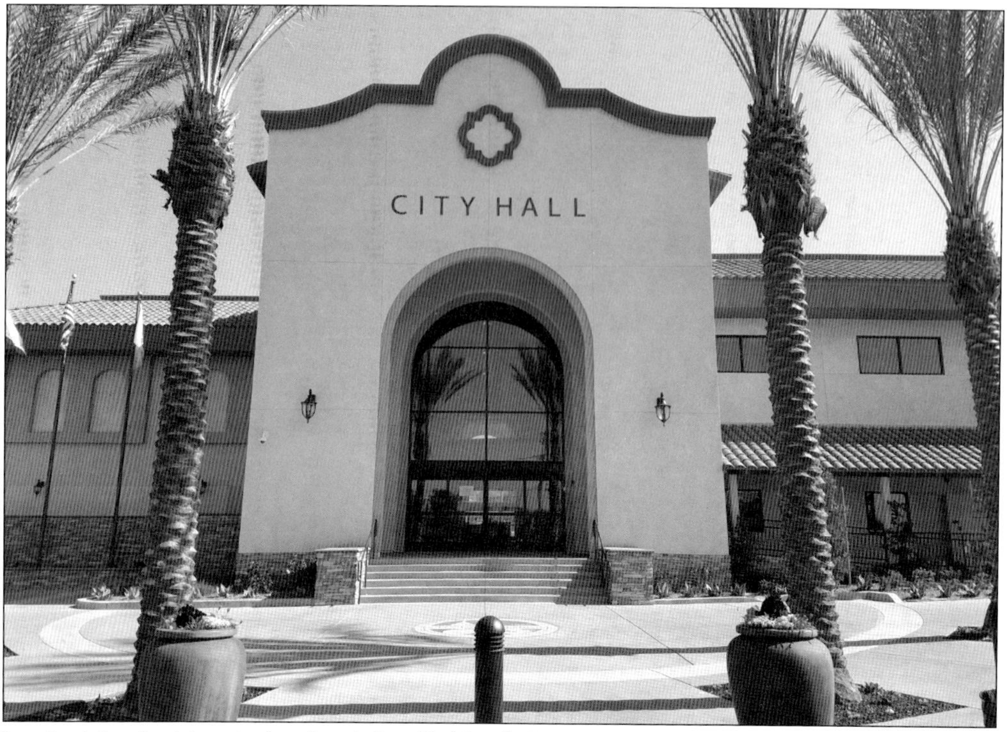

Inspired by the historic facade of the Children's Museum at La Habra's Union Pacific depot building, the new city hall was completed in March 2017. The centerpiece of the new complex is the 5,000-square-foot atrium, which provides the building with lots of natural light and a welcoming mountain lodge feel. In addition to government offices, meeting places, and the city council chambers, the post office also moved across the street to the new building. Many local officials attended the ribbon-cutting ceremony as documented in the picture below, including Mayor Rose Espinoza, Mayor Pro-Tem Tim Shaw, and Council Members James Gomez, Tom Beamish, and Michael Blazey. (Both, courtesy of the City of La Habra.)

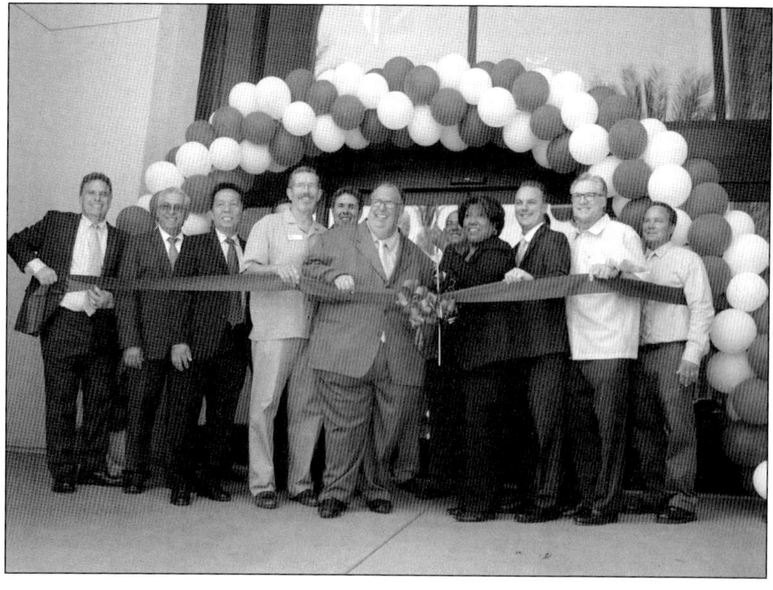

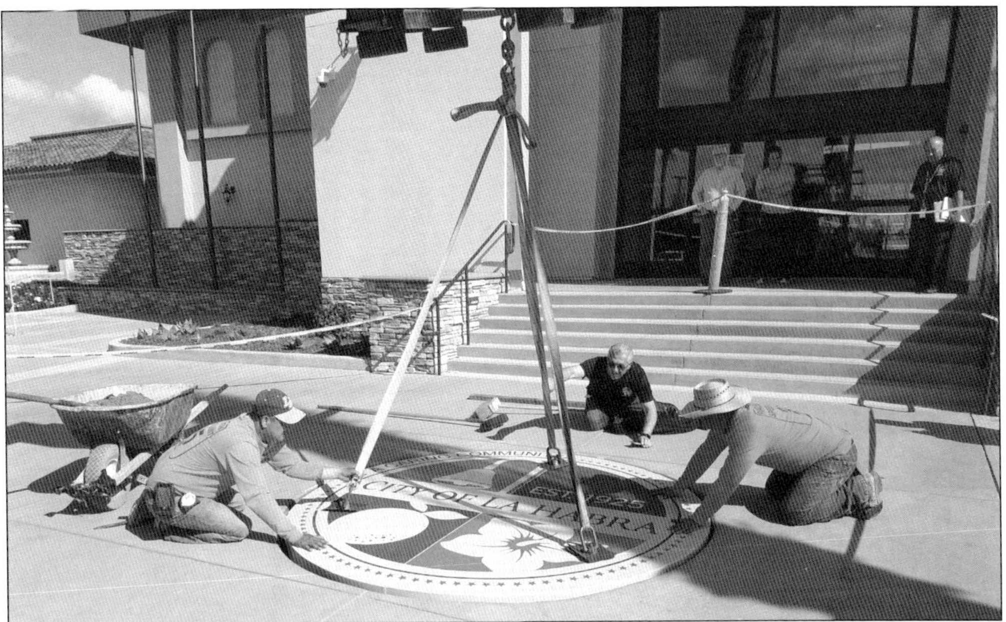

This photograph captures the moment the city seal was placed in front of the new city hall building constructed to replace the former government building that opened in 1969. The project was largely funded through the sale of the old city hall to a residential developer, allowing the building to be completed without adding to the city's debt. (Courtesy of the City of La Habra.)

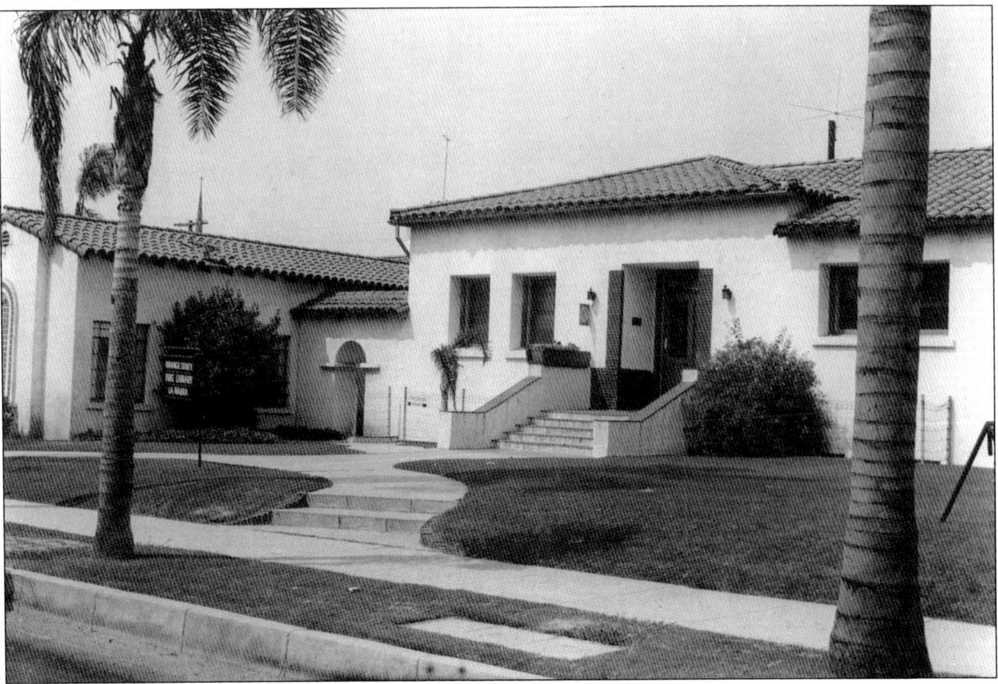

The La Habra Branch of the Orange County Library, seen here in 1957, began with 50 books in a room at the Citizens Bank in the early 1920s and moved into this building within the old civic center complex in 1937. It was replaced with the current library building just a stone's throw away from its predecessor in 1966. (Courtesy of the La Habra Historical Society.)

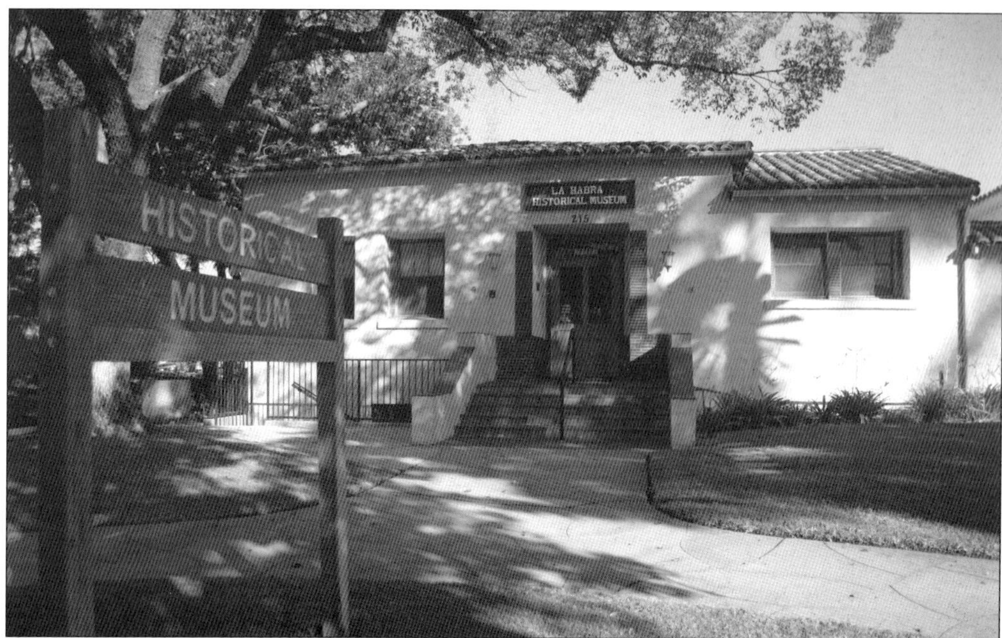

Housed in the former library building in the first civic center, the La Habra Historical Museum was the inspiration and dream of some of La Habra's earliest settlers, including the Sansinena, Smith, Launer, Steele, Ridgway, Proud, and Jenson families. It was founded by the La Habra Old Settlers Historical Society and is directed by Roy Ramsland, president of the Old Settlers. (Courtesy of the City of La Habra.)

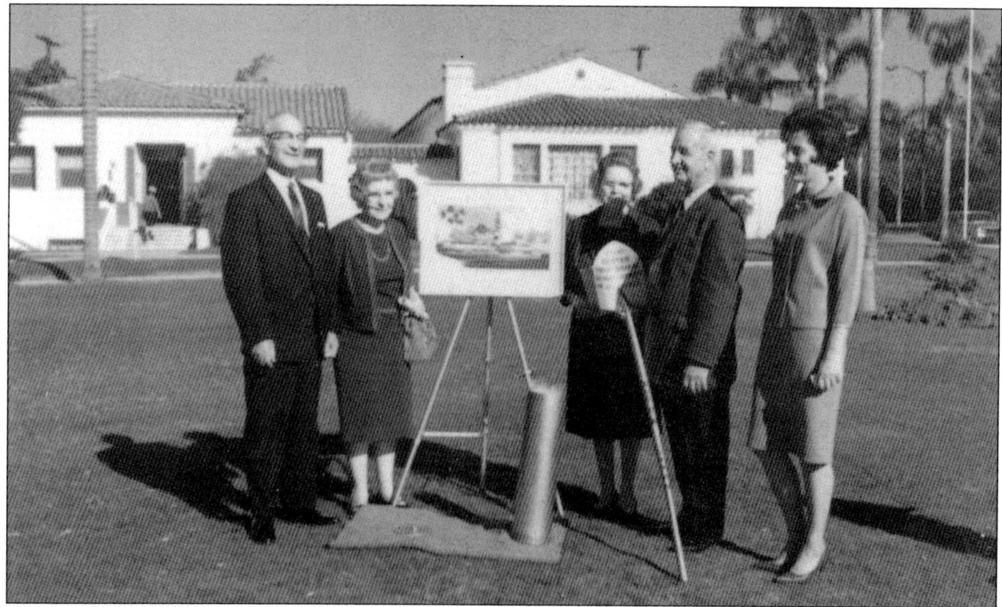

Pictured at the ground-breaking ceremony for the new La Habra Library in 1966 are five unidentified city and county officials. The old civic center, which housed the library until the current building was constructed in the 1960s, can be seen in the background. In 1983, an addition was built onto the existing structure, completing the complex as La Habra residents know it today. (Courtesy of the La Habra Historical Society.)

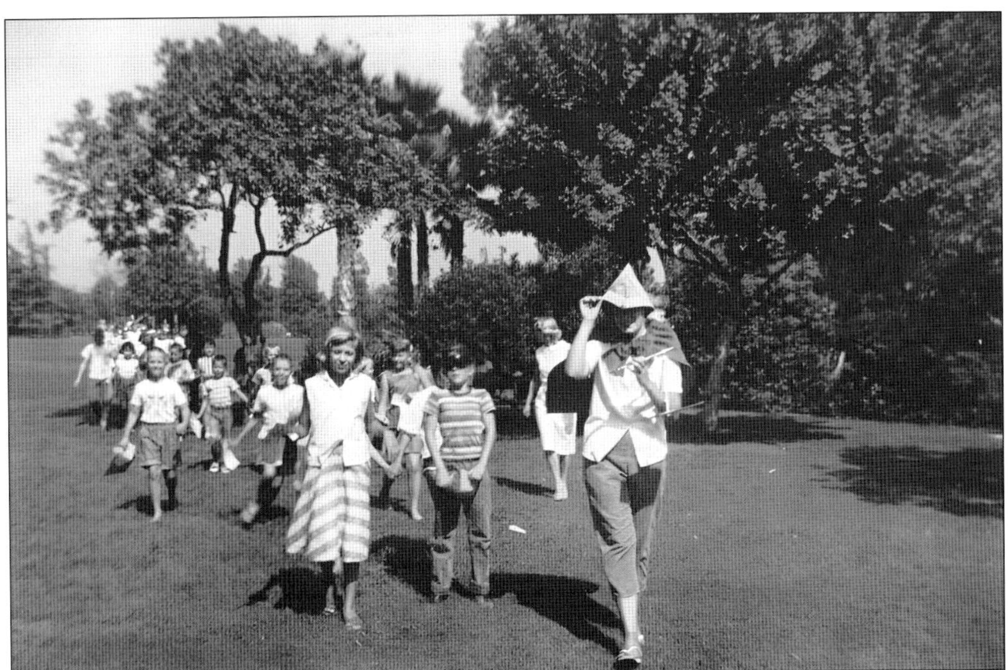

Inspiring children to read has been a key goal of the La Habra Library since early on in its history. In the photograph above, a librarian leads children to the La Habra Library's Summer Reading Party in 1960. According to a note on the back of the photograph, juniors from La Habra High assisted the librarians in putting on this event. In 1983, when an addition was made to the building, a children's listening station was also introduced to the kids' corner of the library. In this area, audio cassette players were built into two large stuffed bears named Bearnard and Bearonica, as seen in the photograph below. (Both, courtesy of Orange County Public Libraries.)

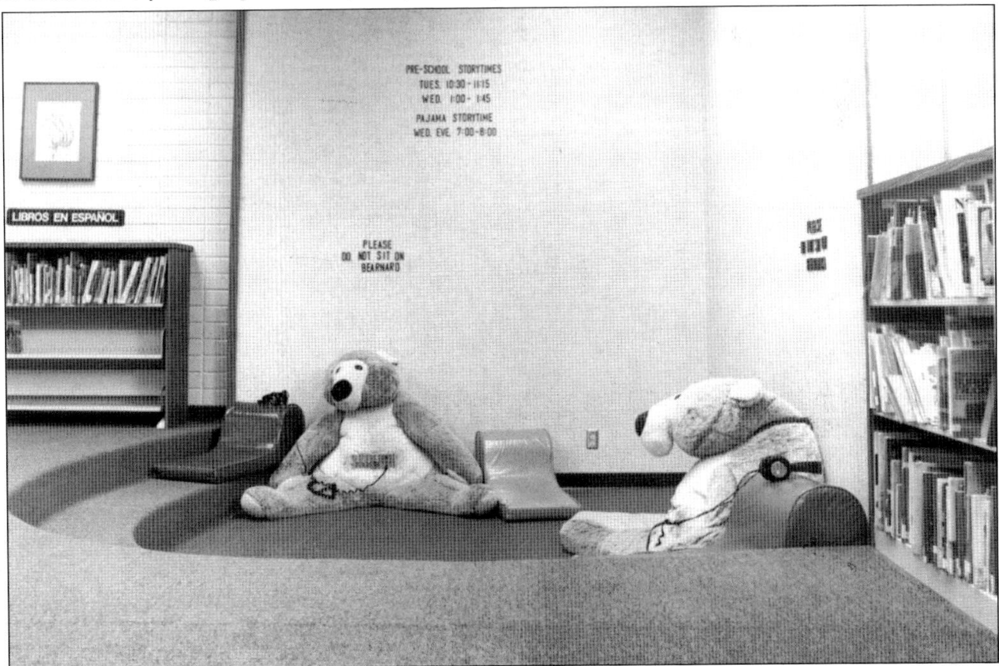

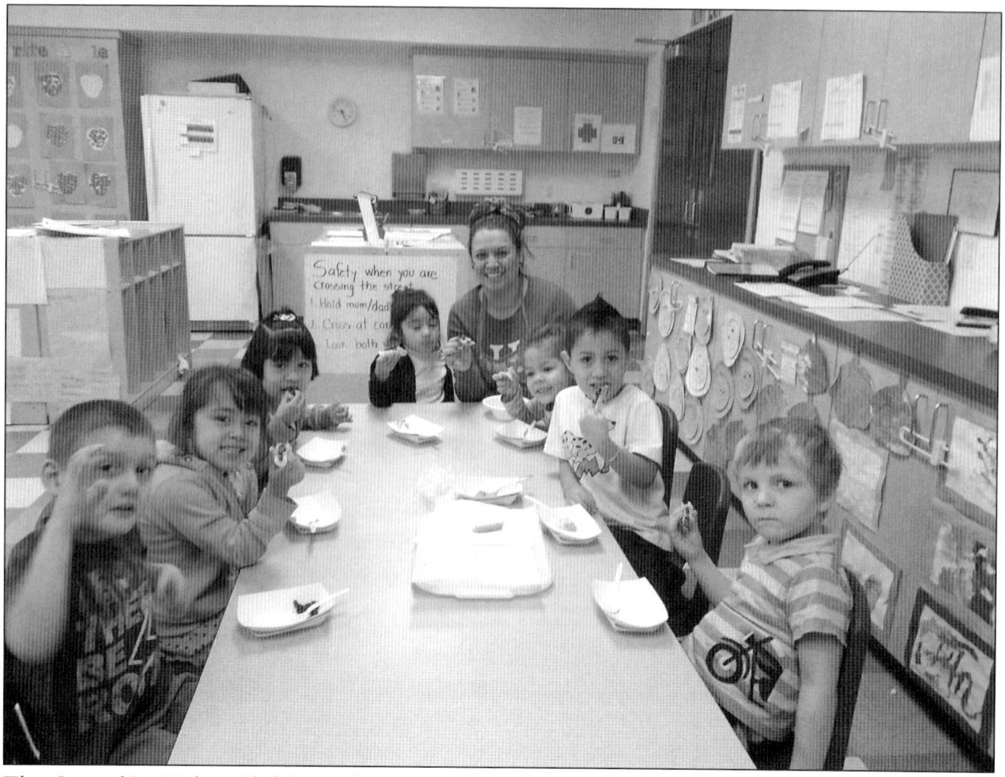

The City of La Habra Child Development Division consists of five programs—Early Head Start, State Preschool, School-Age, Family Child Care, and Family Child Care Home Food Program (CACFP) Sponsor—with the mission to provide quality childcare and development services to children ages 0 weeks to 11 years old, including children with special needs and pregnant moms. (Courtesy of the City of La Habra.)

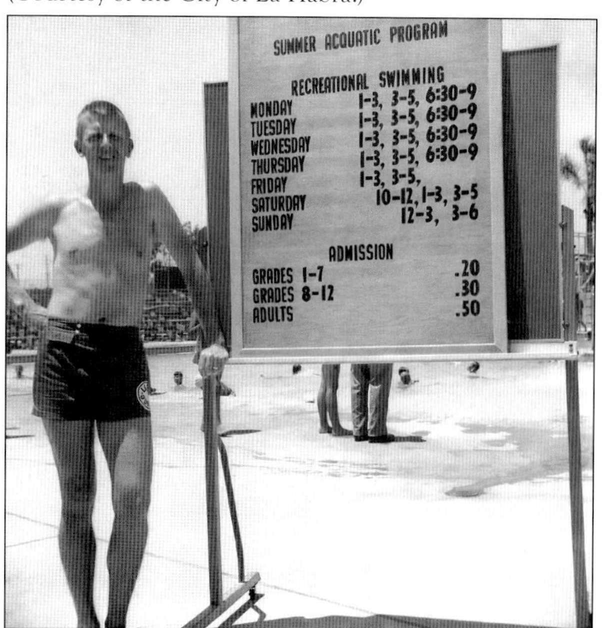

This photograph from the 1960s shows the Summer Aquatics Program hours and admission costs for that season. Then, as today, the city's aquatic programs were offered at La Habra High School's swimming pool, and while the cost of admission might be a bit lower than it is now in the 2020s, the city now has more activities to offer. (Courtesy of the City of La Habra.)

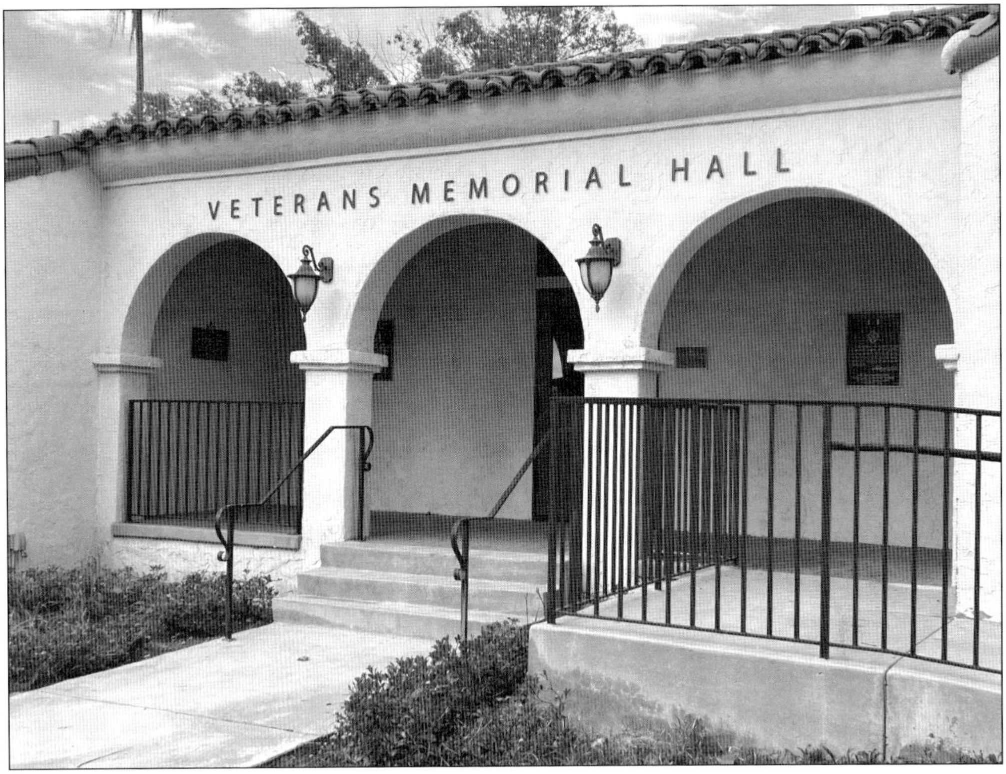

The American Legion post was established in La Habra in December 1922 by veterans of World War I. Sponsoring recreation activities and patriotic meetings, the group raised enough funds to build a large hall in the La Habra Civic Center in the 1930s. The La Habra post is still active in the community today and continues to meet in this historic building. (Courtesy of the City of La Habra.)

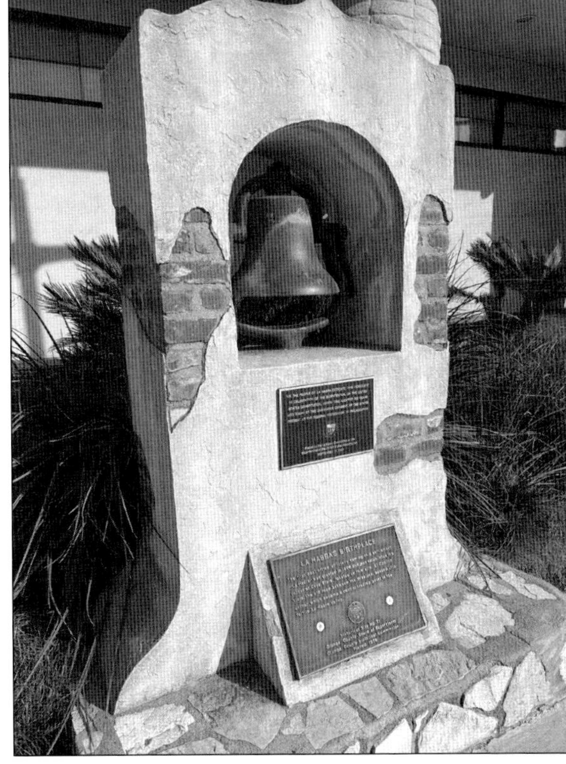

This bell monument honors "the birthplace of La Habra," the corner of La Habra Boulevard and Euclid Street. The intersection received the nickname when post office documents arrived at the post office here inside Zachary Coy's store on December 21, 1895, finally making the name "La Habra" official for the first time. (Courtesy of the City of La Habra.)

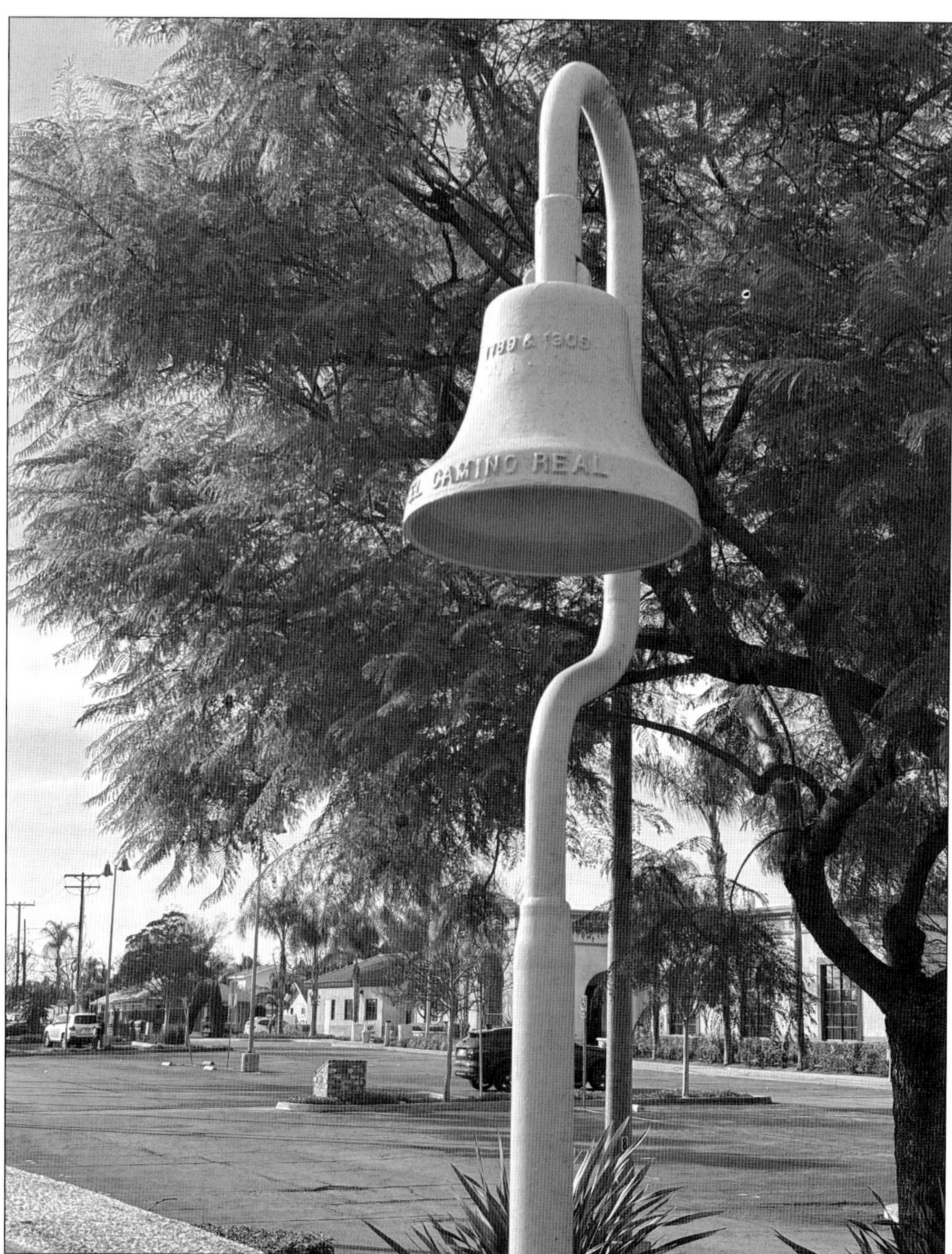

A 1904 Los Angeles convention aimed to place commemorative bells along El Camino Real, a network of trails connecting the Spanish missions in California during the 18th and 19th centuries, to mark this historic route. The La Habra Woman's Club selected Harbor and La Habra Boulevards for the city's first bell in 1908, as the route once passed over what is now known as La Habra Boulevard, but it was destroyed by runaway horses that same year. Much later, present-day replica bells were placed at significant historical sites along La Habra Boulevard, such as the one seen here. All bell locations were originally designed to stand one mile apart from each other, but over the century, some have been damaged or stolen. (Courtesy of the City of La Habra.)

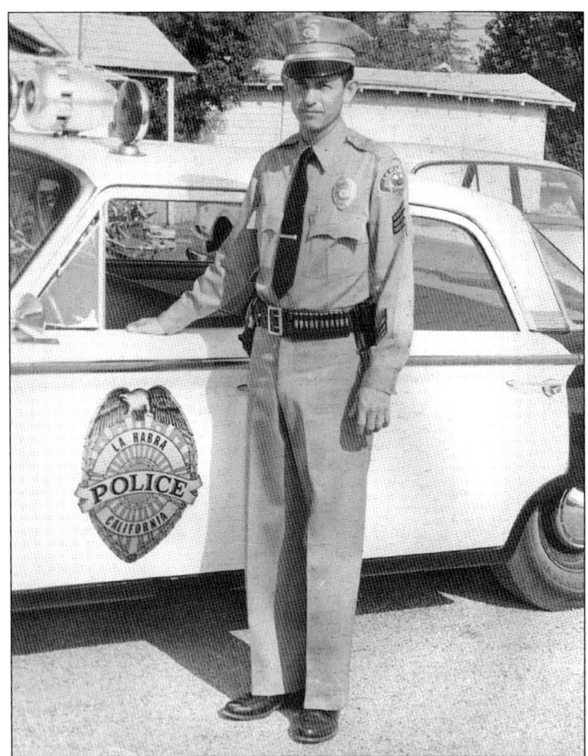

A La Habra police officer poses with his patrol car in the 1950s. The force has grown from its modest beginnings, now boasting 27 officers, eight sergeants, and two lieutenants assigned to patrol. One such lieutenant was Mel Ruiz, who began his career with the department in 1983 as a volunteer police explorer; he was hired in part-time positions before becoming a full-time police officer in 1987. Among many other roles, he served as the department's first youth service officer, a special investigations unit gang detective, and a professional standards unit lieutenant. Ruiz is credited with bringing the "Coffee with a Cop" and "Cool Cop" events to the city. He retired in 2017 after 30 years with the department but continues to serve the city as a part-time reserve detective. (Right, courtesy of the City of La Habra; below, courtesy of Wendy Guandique.)

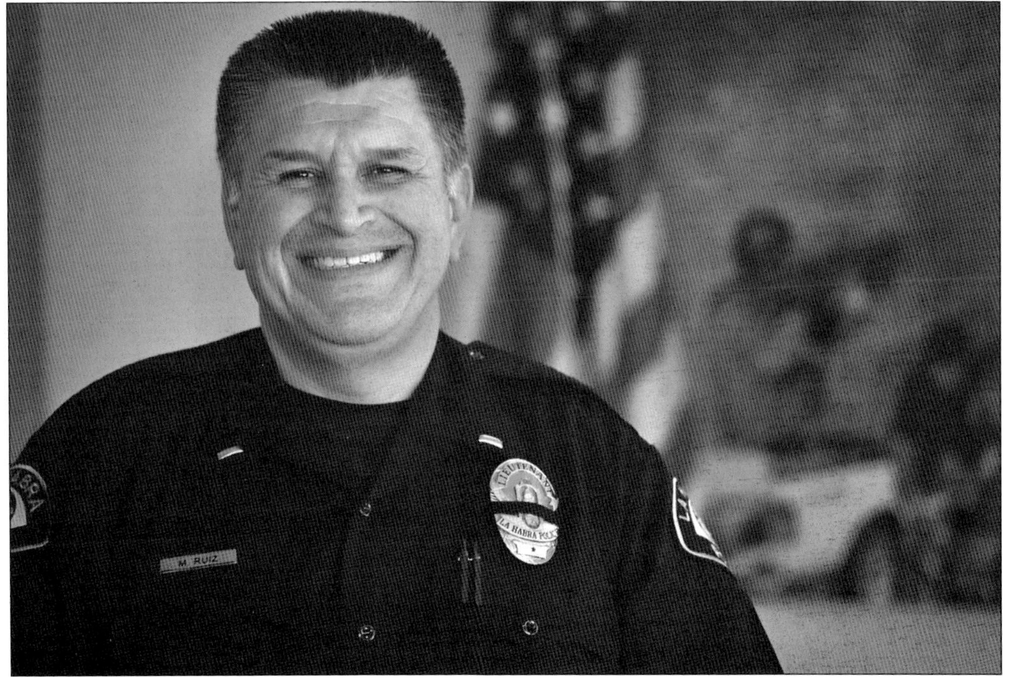

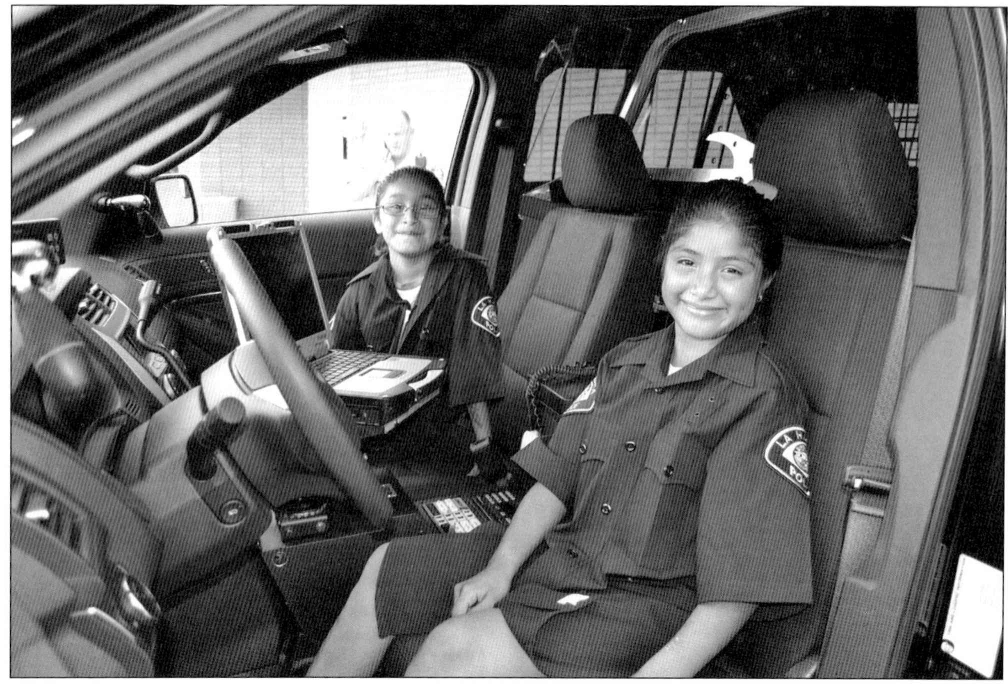

Two young La Habrans pose in a patrol car and police uniforms at the police department's open house in October 2024. The open house, as well as the La Habra Police–sponsored National Night Out, are designed to strengthen the community by encouraging citizens to engage in stronger relationships with each other and local law enforcement partners. Below, chaplain Mike Murphy and his canine partner, Ike, greet children and families at that year's National Night Out. Chaplain Murphy and Ike provide support in traumatic situations that can arise in police work. (Above, courtesy of David DeLeon; below, courtesy of the City of La Habra.)

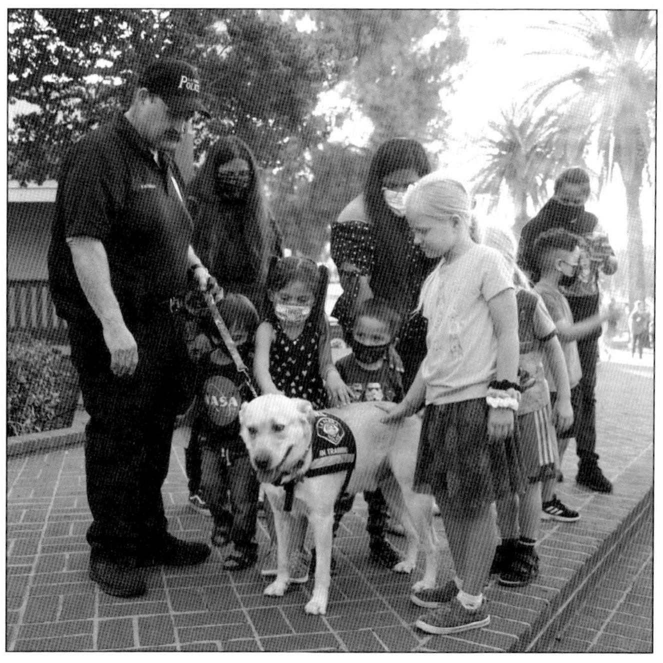

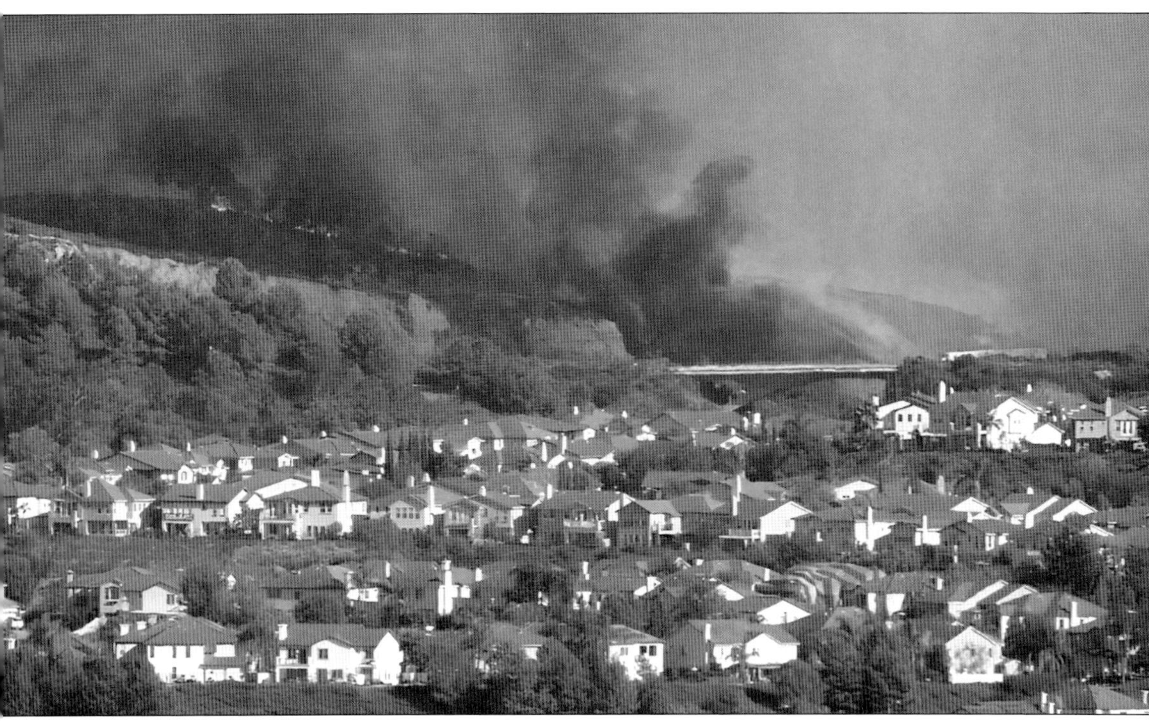

This historic fire in Los Coyote Hills threatened the homes of La Habra residents in 2015 and spread up to 80 acres. Los Angeles County firefighters saved hundreds of homes in their collaboration with the Orange County Fire Authority and two Los Angeles County Fire "super scooper" planes. (Courtesy of David DeLeon.)

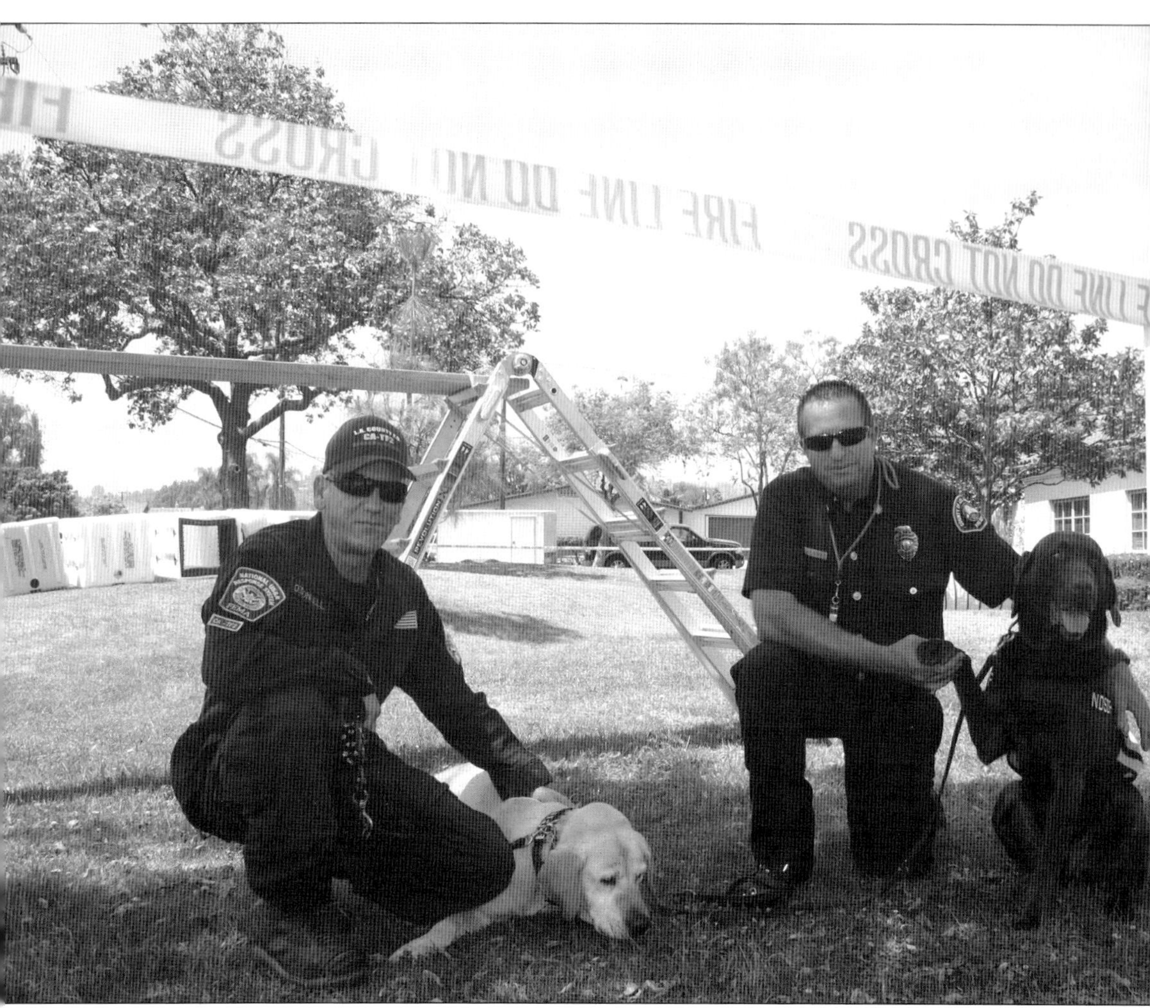

Since 2005, the City of La Habra has had contracts with the Los Angeles County Fire Department, one of the largest and most sophisticated fire departments in the United States, to provide general fire and emergency medical services. Los Angeles firefighters pose with the department's rescue dogs in this photograph from Fire Service Day in 2015, an event for residents to meet firefighters, watch demonstrations, and learn about fire safety. (Courtesy of the City of La Habra.)

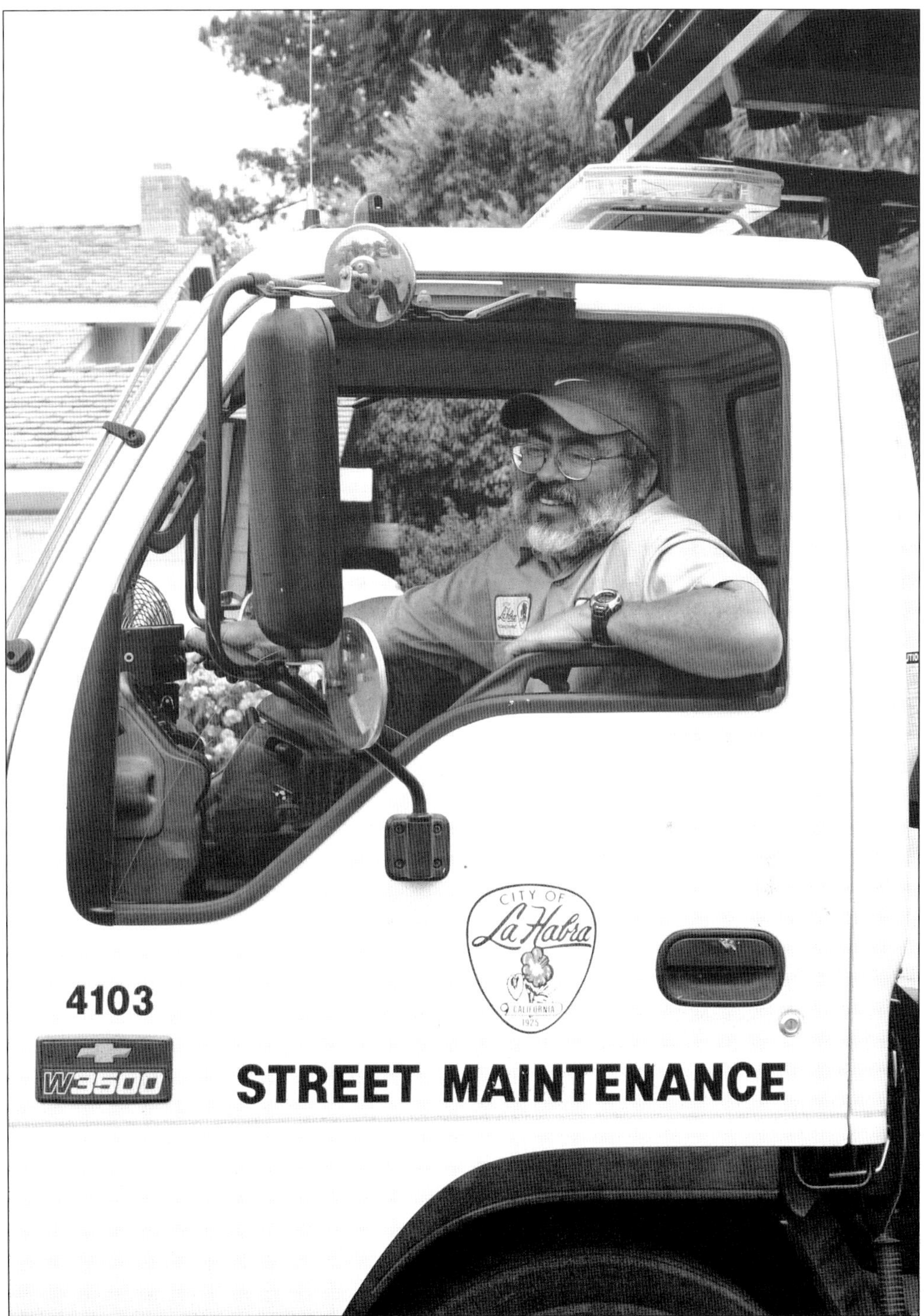

The Department of Public Works is responsible for developing and maintaining the city's infrastructure through its seven divisions, which include administration, engineering, fleet maintenance, parks and trees, refuse, street maintenance, and water/sewer. This photograph of one of the department's dedicated public workers is from 2007. (Courtesy of the City of La Habra.)

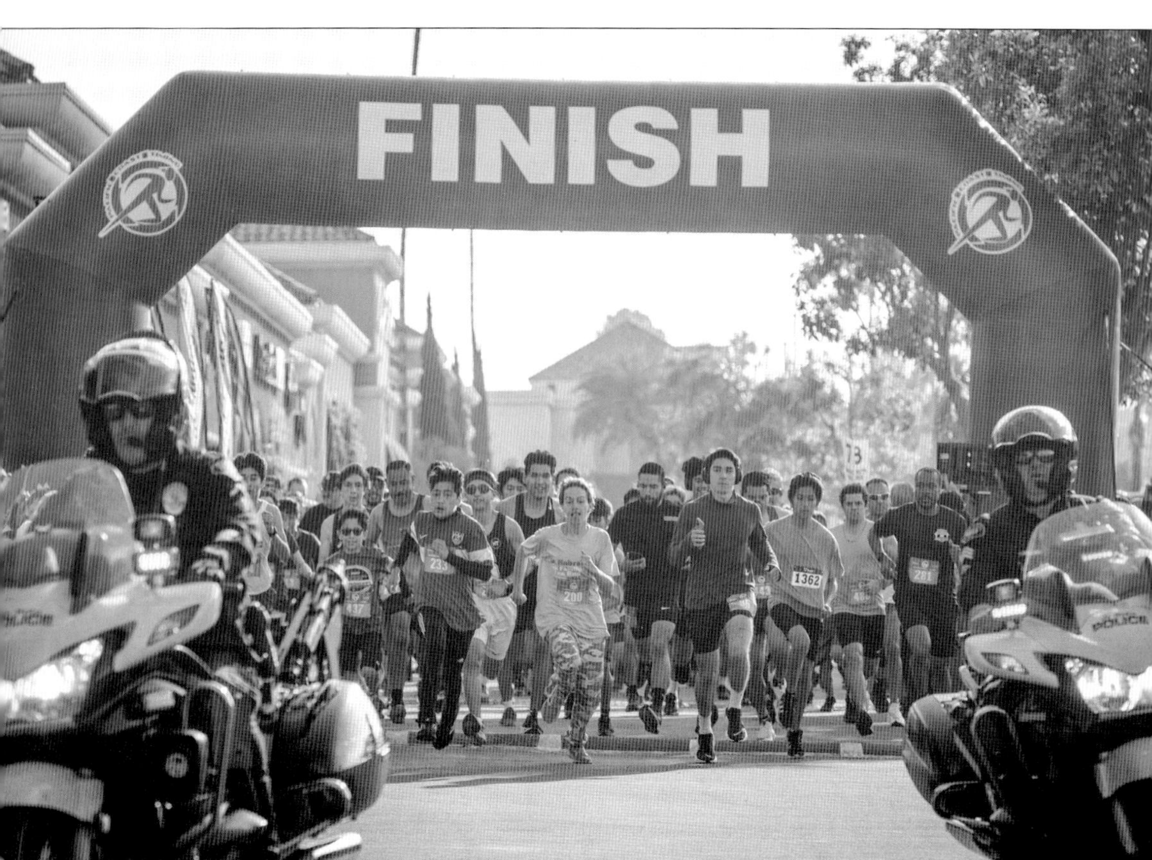

Each year, the city organizes the La Habra Races, which include a 5K run/walk, a 1K "kiddy run," and a 50-yard "diaper dash" with a different theme each year. In this photograph from 2024, motorcycle officers from the La Habra Police Department escort the young runners to the finish line. (Courtesy of the City of La Habra.)

Ten

Memorable Moments

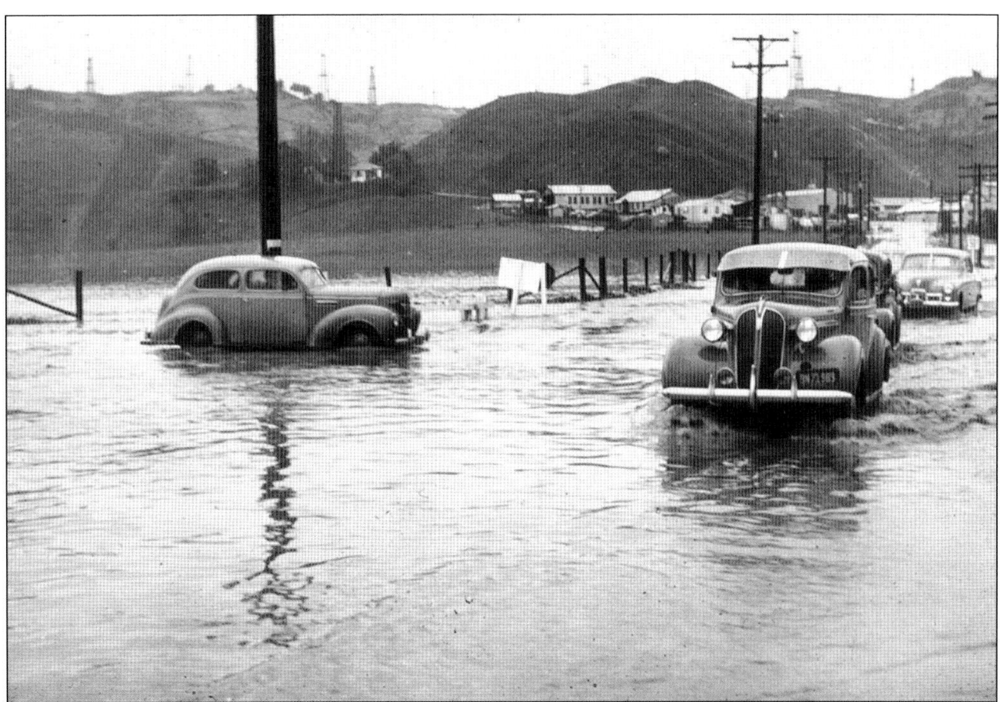

With development well underway in La Habra, the Los Angeles River Flood of February–March 1938 hindered and demolished much of the city's progress. The flood, one of the most disastrous in the history of Southern California, generated almost one year's worth of rainfall in just a few days. Here, cars can be seen navigating through the aftermath at Beach Boulevard and Imperial Highway. (Courtesy of La Habra Historical Society.)

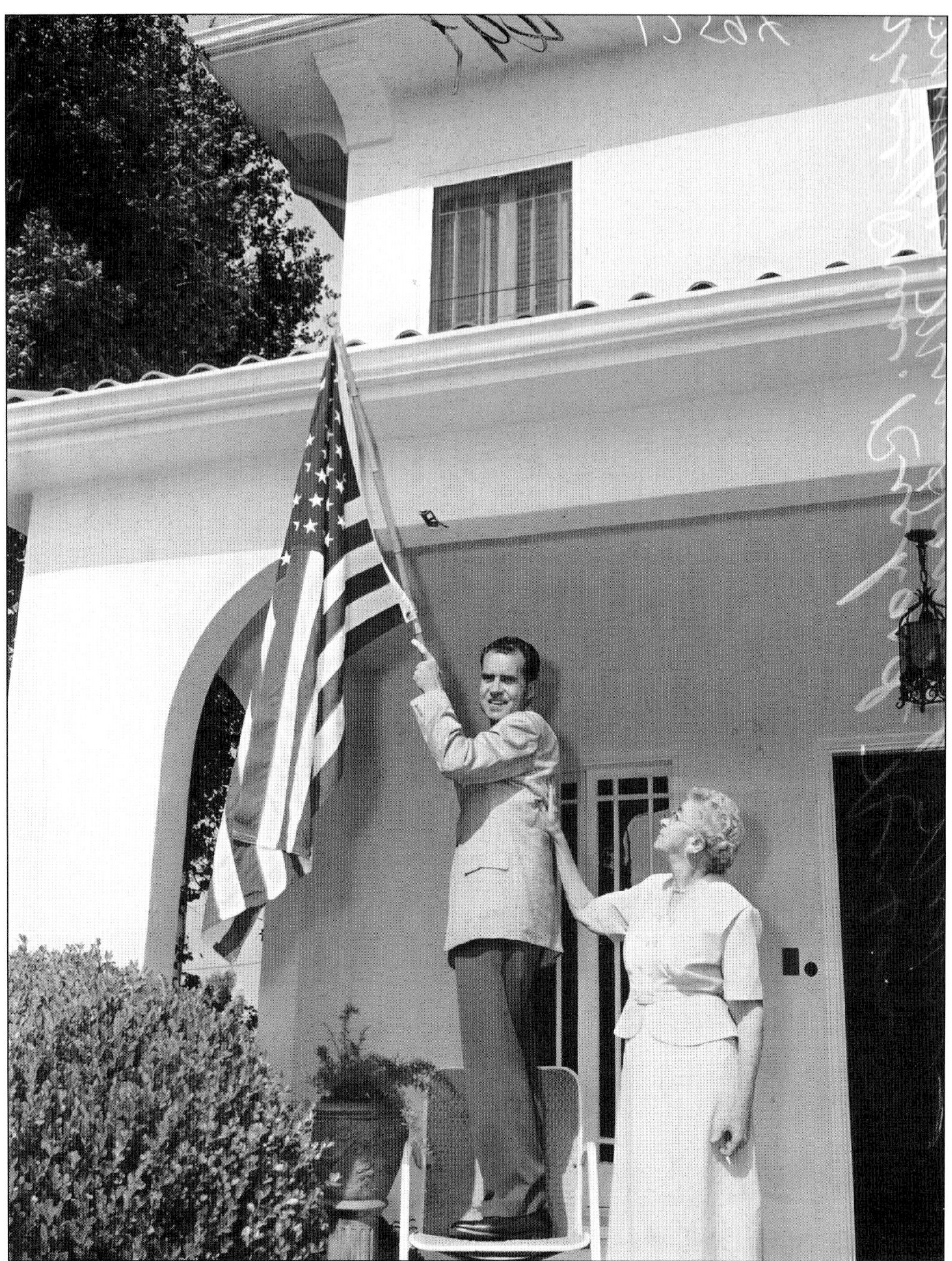

Descending from early La Habra settlers, Pres. Richard Nixon hangs an American flag at his mother's La Habra home. Nixon opened his first law office on La Habra Boulevard in August 1939 and left the firm in January 1942. The building was demolished in October 1992, but a commemorative plaque constructed of the structure's original bricks stands in its place. (Courtesy of the City of La Habra.)

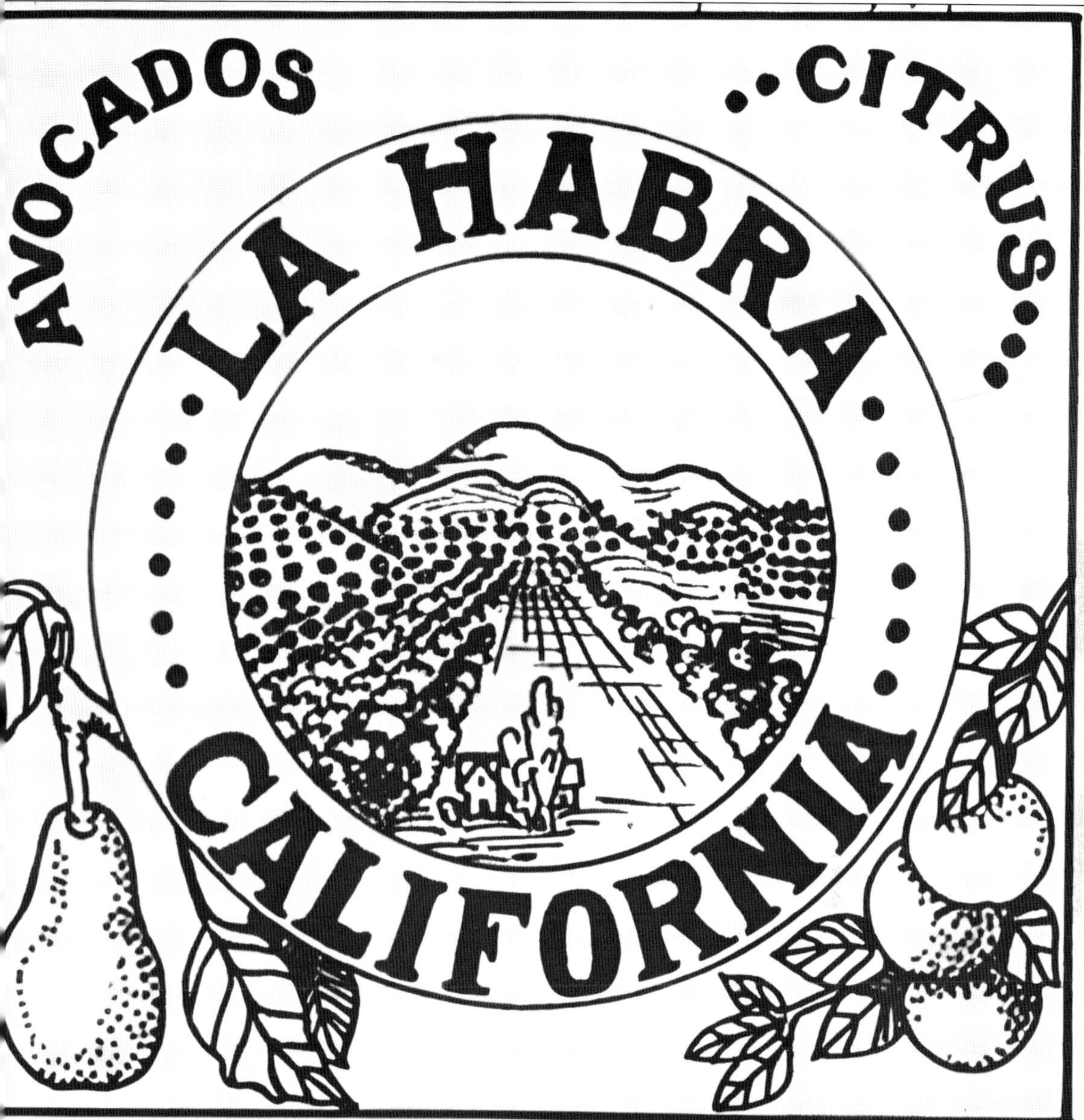

To commemorate the 20th anniversary of the US airmail service and promote its usage, Postmaster General James A. Farley and Pres. Franklin Roosevelt created Air-Mail Week. Many small towns rallied behind the event and created their own designs to display on packages that were posted during the week-long celebration, including La Habra, whose special stamp highlighted the city's flourishing agricultural community. (Courtesy of the La Habra Chamber of Commerce.)

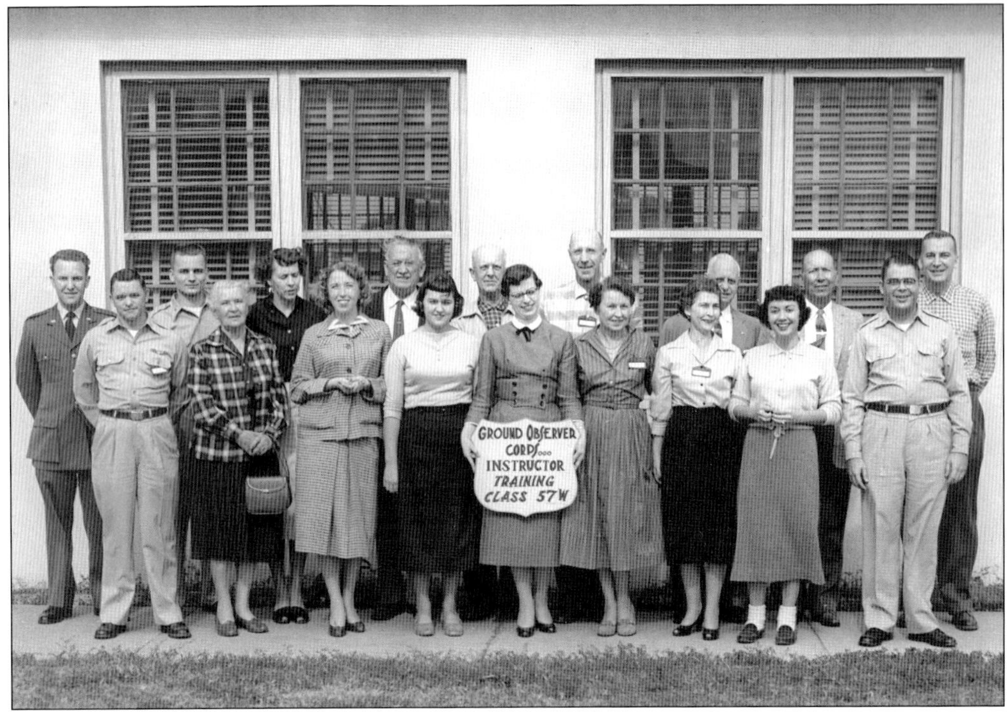

During World War II and in the first years of the Cold War, many cities in California and other coastal states relied on Ground Observer Corps like this one from La Habra pictured during their instructor training in 1957. Equipped only with binoculars, wall charts, and model airplanes, these volunteers, typically housewives and teenagers, kept careful eyes on the skies to determine if the planes that flew overhead were American or enemy and plot their flight path. In La Habra, the watchtower the volunteers used was atop the police station, as pictured below. (Both, courtesy of the La Habra Historical Society.)

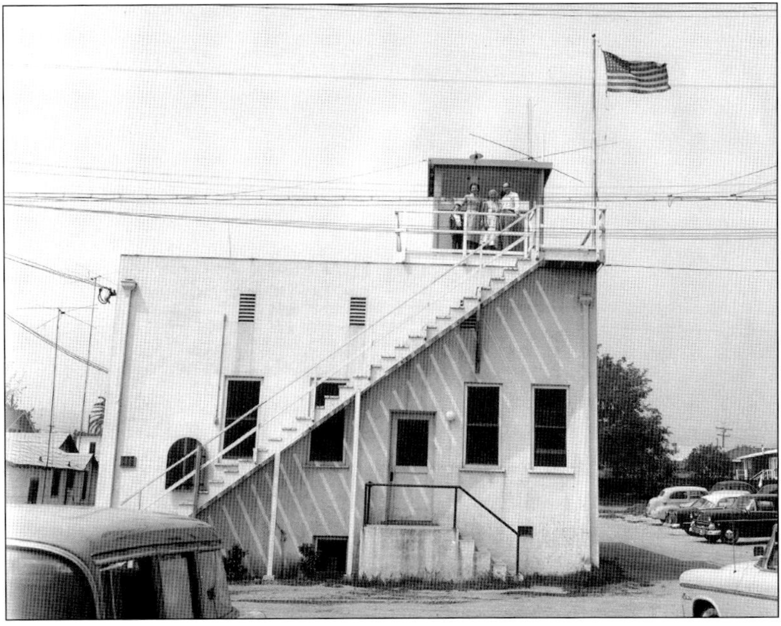

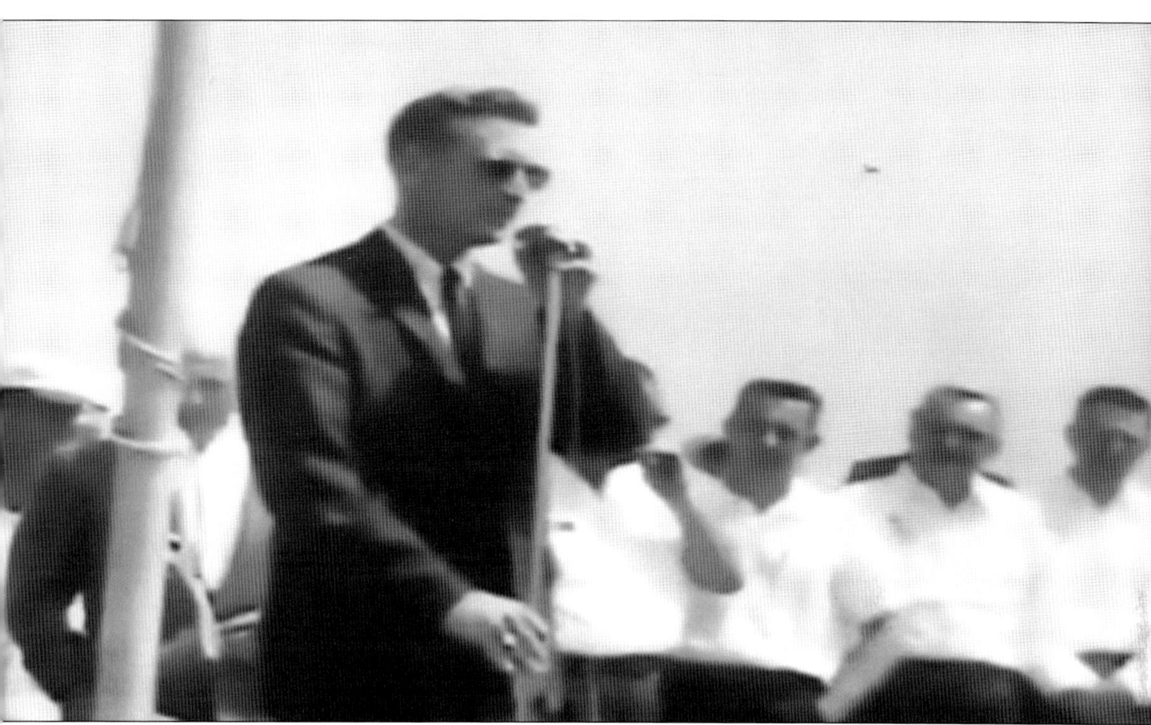

This image, taken from Super 8 footage of an opening-day ceremony for La Habra's Little League Baseball in 1960, documents a special guest appearance by Vin Scully, the longtime announcer for the Los Angeles Dodgers, shortly after the team made the move from Brooklyn. The baseball field, which once sat on what many know as Vista Grande Park, was dubbed "Scully Stadium" in honor of the beloved sportscaster's visit to La Habra. The fields of Scully Stadium were shut down in the 1980s due to the land settling under the former landfill and remained undeveloped for many years. However, the city has begun construction on a new park, which will be known as Vin Scully Centennial Park and is set to contain the city's first dog park. (Courtesy of James Gomez.)

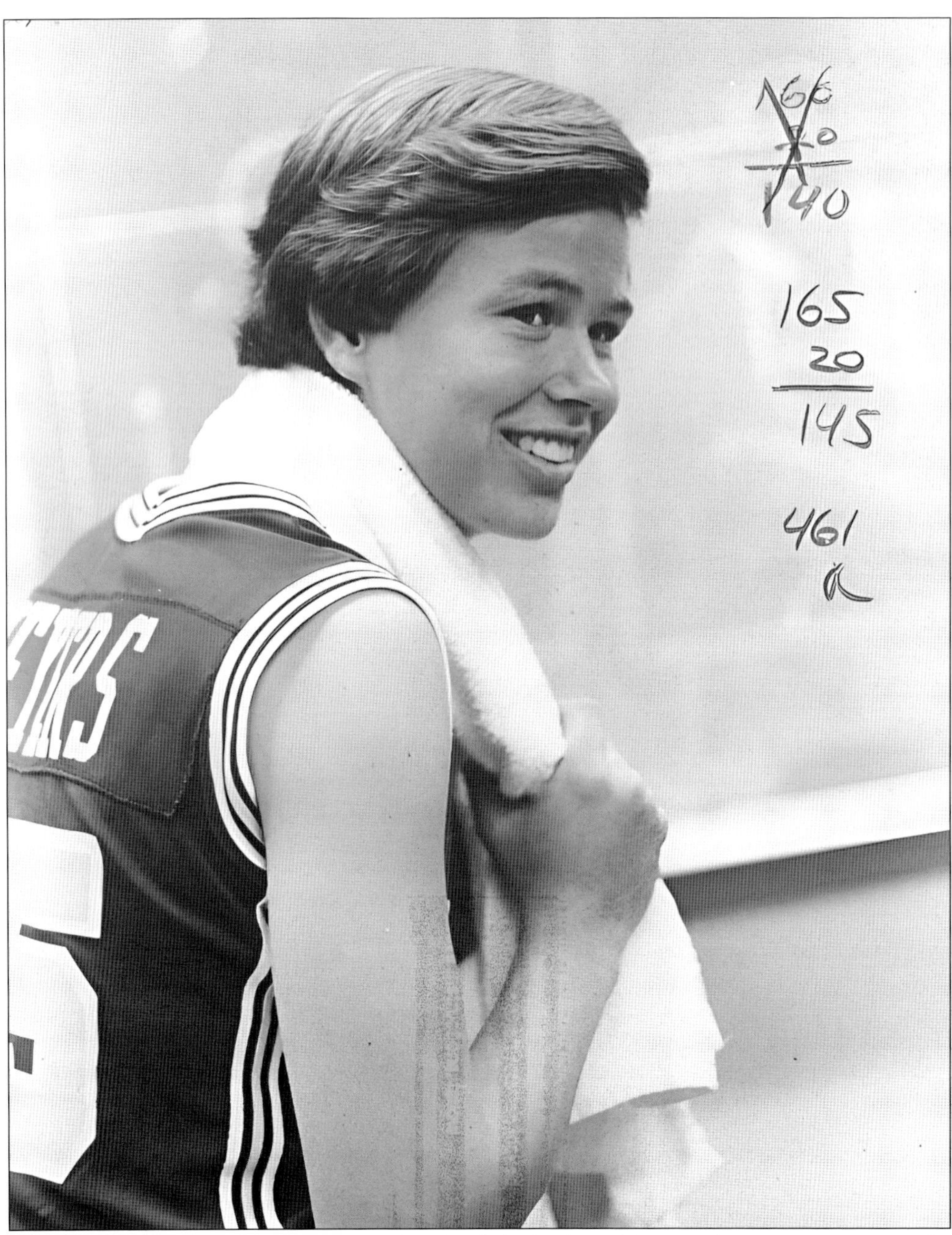

Ann Meyers, a graduate of Sonora High School, had a lot of firsts in her career. In 1974, she became the first high school student to play for the US National Basketball Team. She was the first women's college basketball player to earn a four-year scholarship. Playing for UCLA, she recorded the first quadruple-double in NCAA Division I basketball history and was the first four-time All-American Women's Basketball Player. In 1979, she was the first woman to sign with an NBA team with the Indiana Pacers. She was an Olympic gold medalist, respected sportscaster, and trailblazer in women's basketball. (Courtesy of the Los Angeles Public Library Photo Collection.)

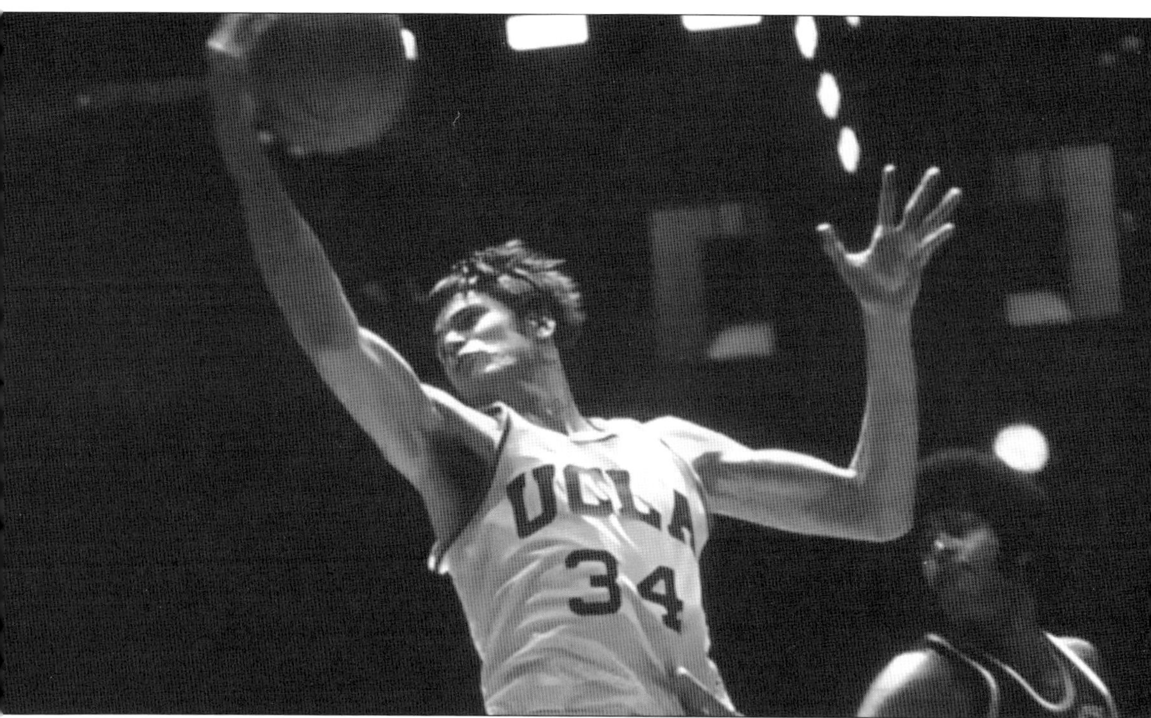

Dave Meyers (right) was the older brother of Ann Meyers and another famous athlete to graduate from Sonora High School. The power forward won two national championships with UCLA in 1973 and 1975. In 1975, he also appeared on the cover of *Sports Illustrated*. He went on to play four seasons with the Milwaukee Bucks. (Courtesy of the UCLA Alumni Association.)

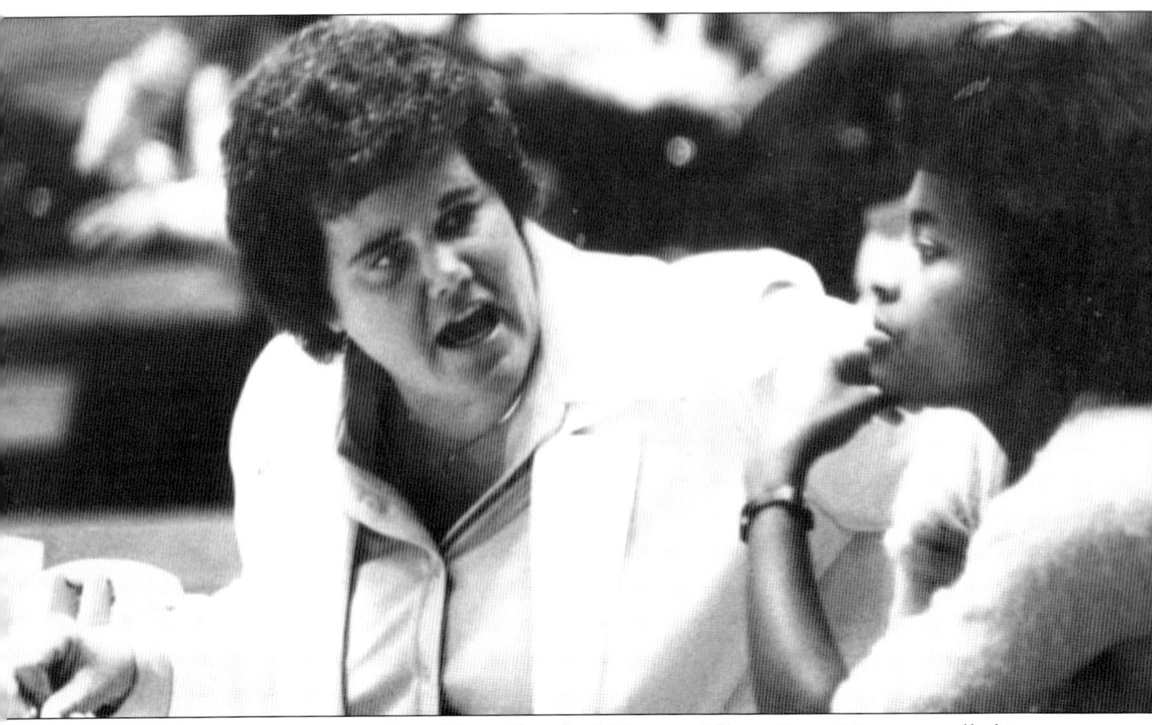

Not to be forgotten among the athletically gifted Meyers siblings, Patty Meyers excelled in many sports at La Habra High School, won several titles at Fullerton College, and was part of Cal State Fullerton's 1970 Association for Intercollegiate Athletics for Women (AIAW) National Championship women's basketball team prior to the institution of the NCAA title tournament. After her playing days were over, she became a successful coach for the Pepperdine women's basketball team. (Courtesy of Pepperdine University Athletics.)

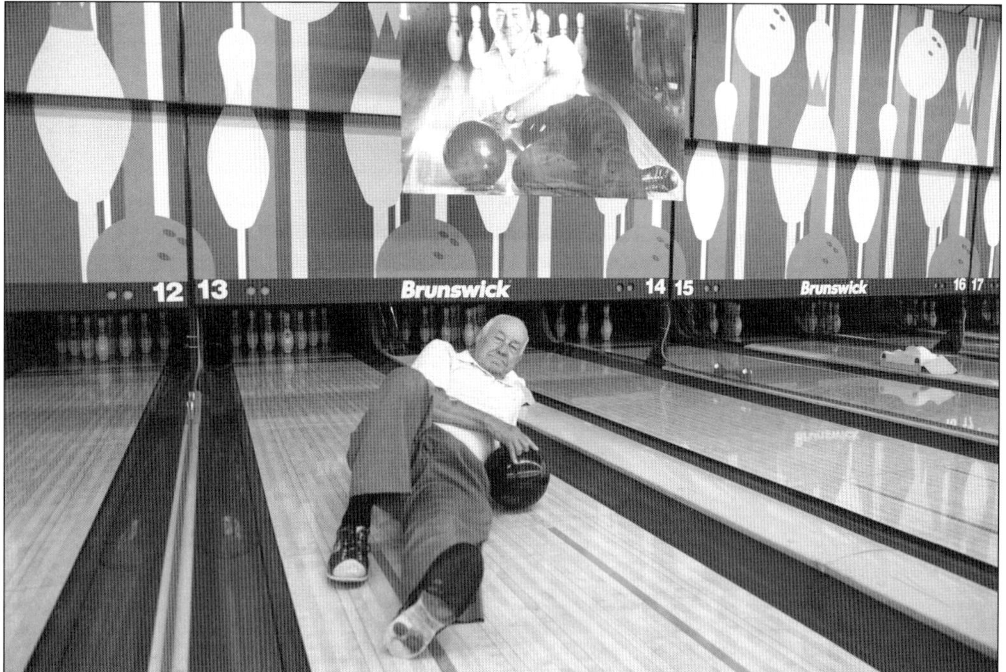

La Habra 300 Bowl is not just famous for its Googie sign and its place in the hearts of many La Habrans. In 1982, it took on new notoriety after professional bowler Glenn Allison shot the first-ever perfect 900 series on lanes 13-14; however, because of the alley not meeting certain lane regulations, this record is controversial to some in the bowling community. Allison, who was born in neighboring Whittier, was also a founding member of the Professional Bowlers Association (PBA) and won five PBA Tour Titles. A wall outside the alley where Allison has estimated he bowled over 10,000 games is dedicated to his historic and contentious accomplishment. (Above, courtesy of ABC News; below, courtesy of La Habra 300 Bowl.)

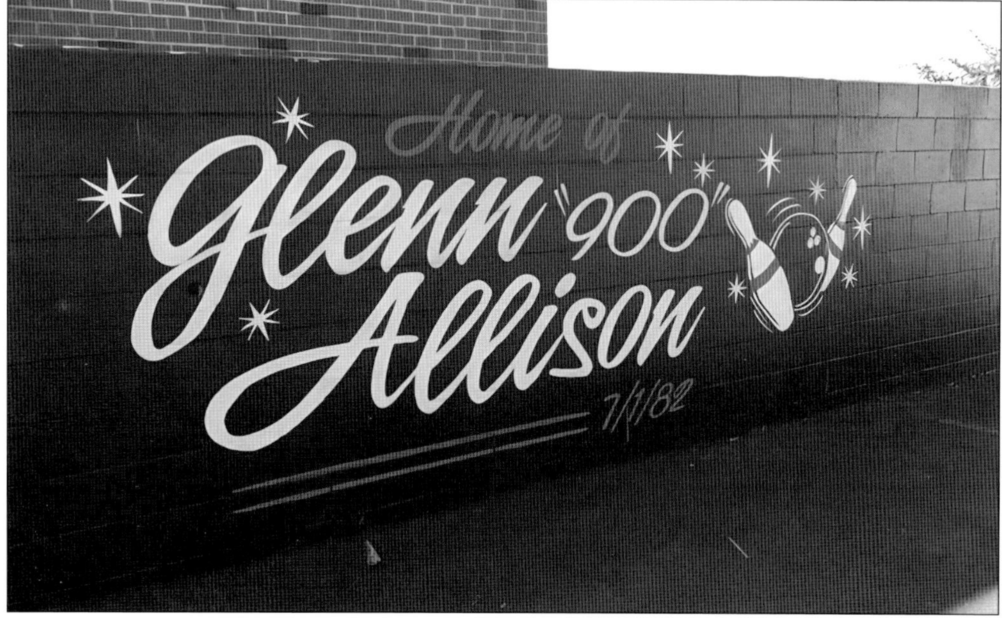

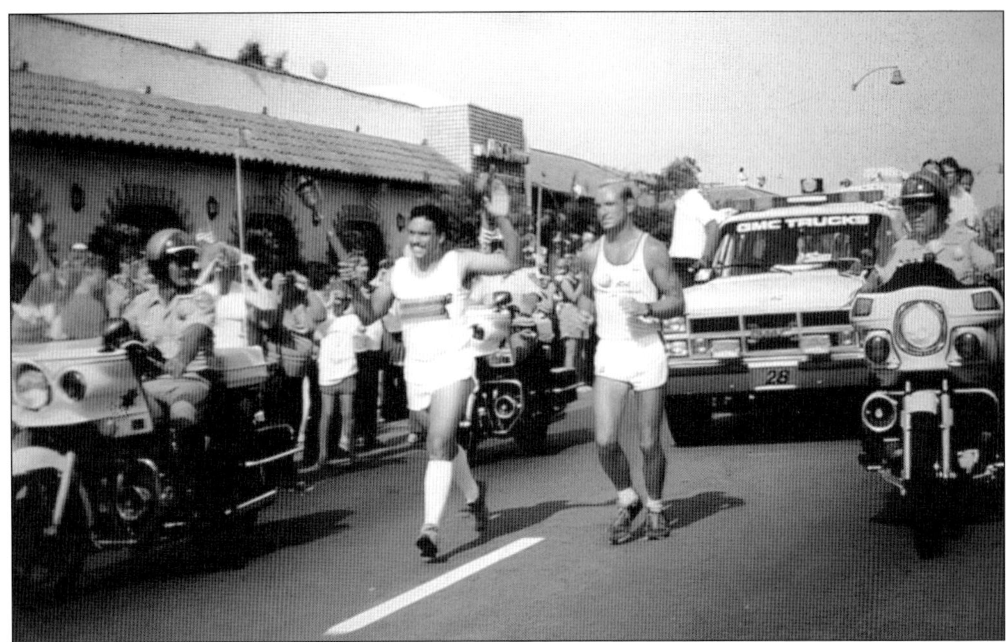

The 1984 Summer Olympics torch relay run passed through La Habra on July 28. The 9,375-mile run from New York City to Los Angeles was by far the longest Olympic torch relay organized up to that point, taking 82 days to complete entirely on foot. The relay was routed through both Harbor and La Habra Boulevards, where onlookers lined the sidewalks to rally in support. This was the first time in Olympic history that members of the general public were permitted to carry the Olympic torch, instead of the previously carefully selected runners, which allowed former mayor and council member Juan Garcia to carry the torch during its leg through La Habra, seen below. (Both, courtesy of the City of La Habra.)

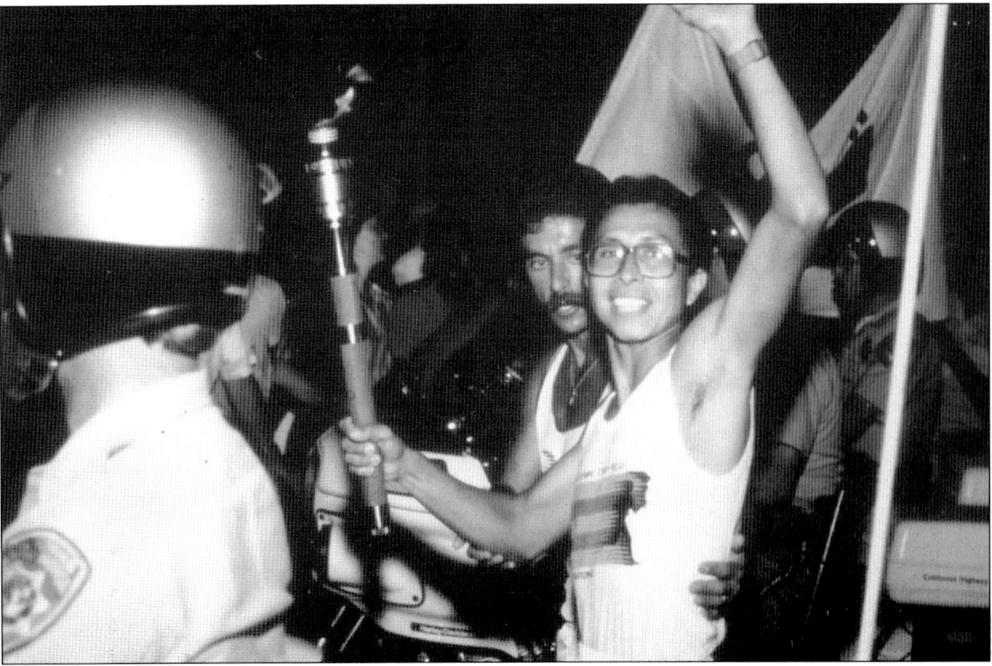

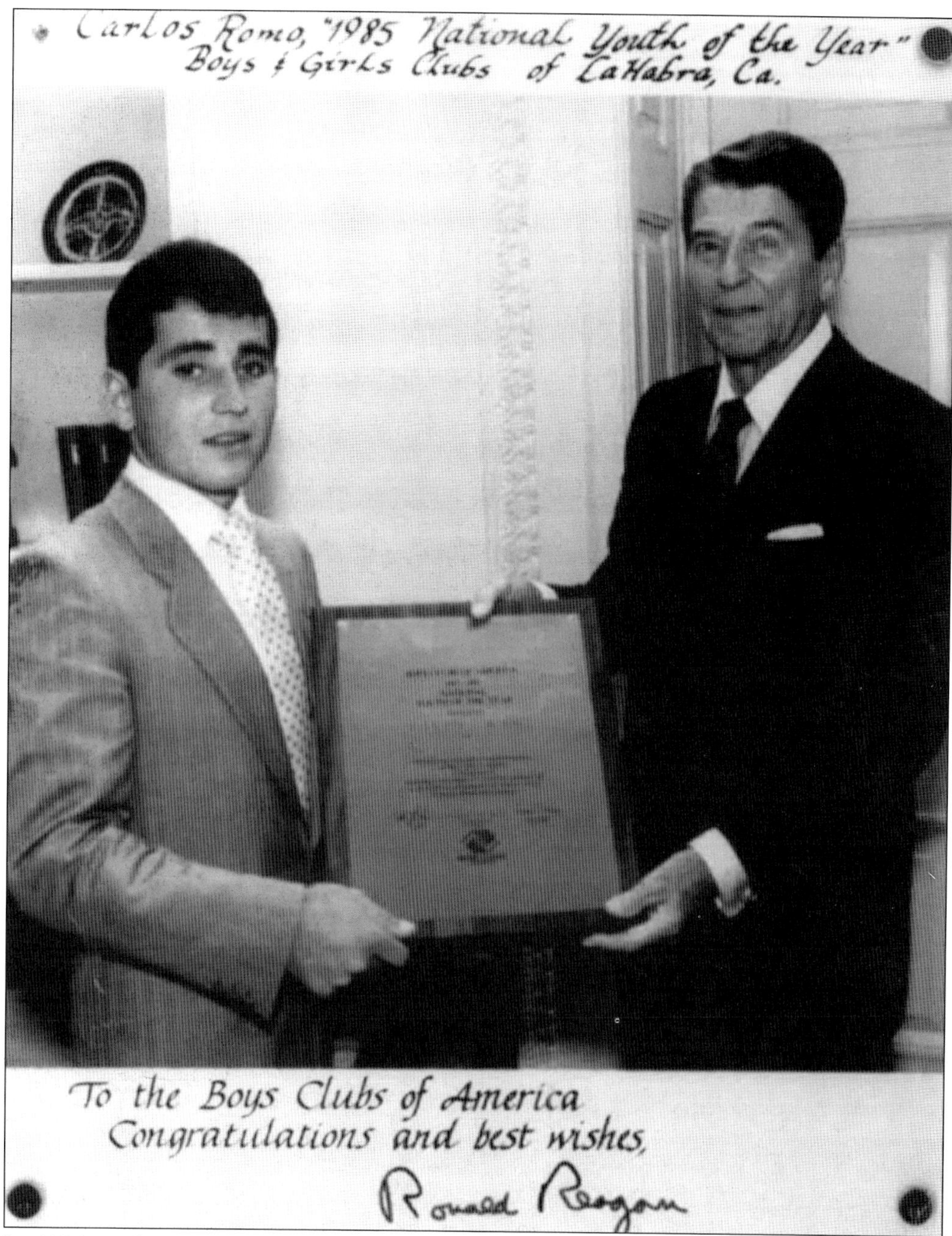

In 1985, Carlos Romo won the prestigious Youth of the Year Award from the Boys Club of America. As part of the award, he traveled to Washington, DC, met with Pres. Ronald Reagan, and got to sit in his chair at the Oval Office. (Courtesy of the Boys & Girls Club of La Habra.)

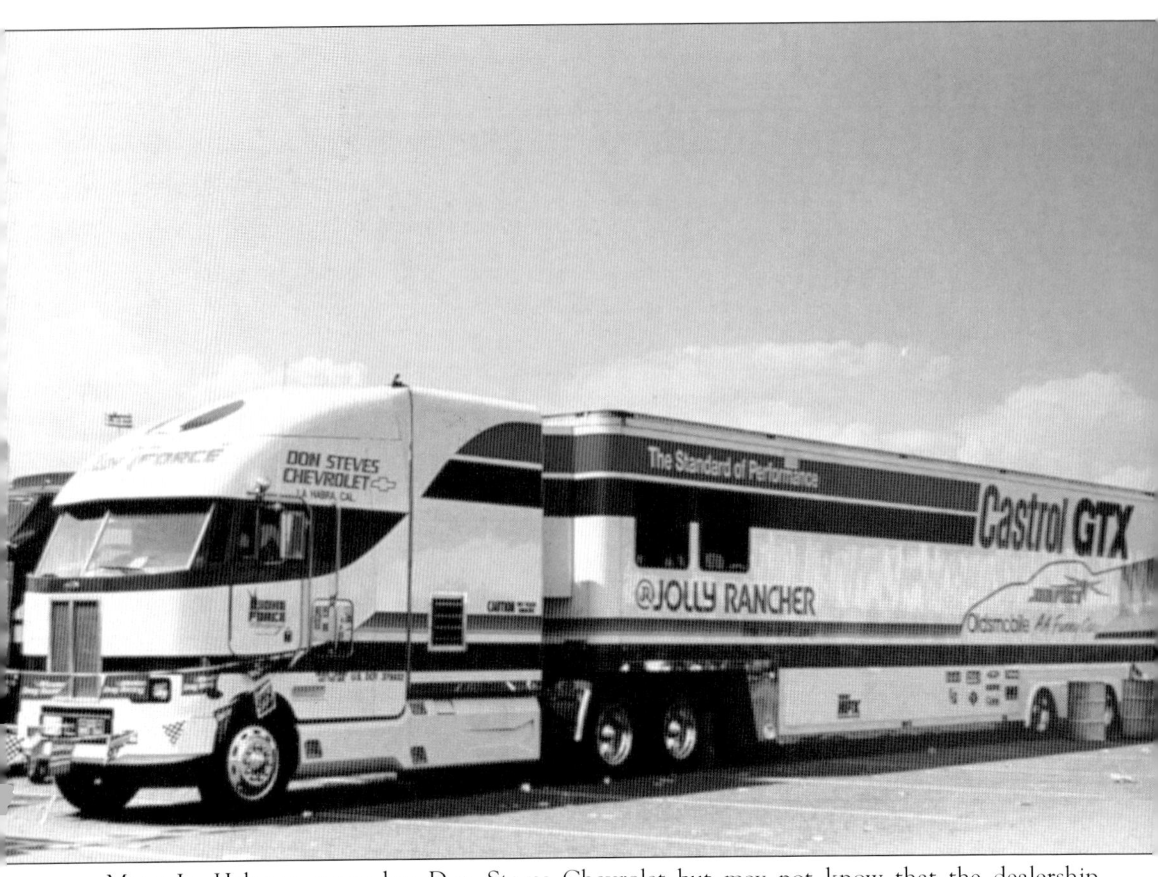

Many La Habrans remember Don Steves Chevrolet but may not know that the dealership sponsored the National Hot Rod Association drag racer John Force. Pictured here is Force's transporter semi with the Don Steves Chevrolet logo on the cab. Force has had 16 championship titles during his career. (Courtesy of the City of La Habra.)

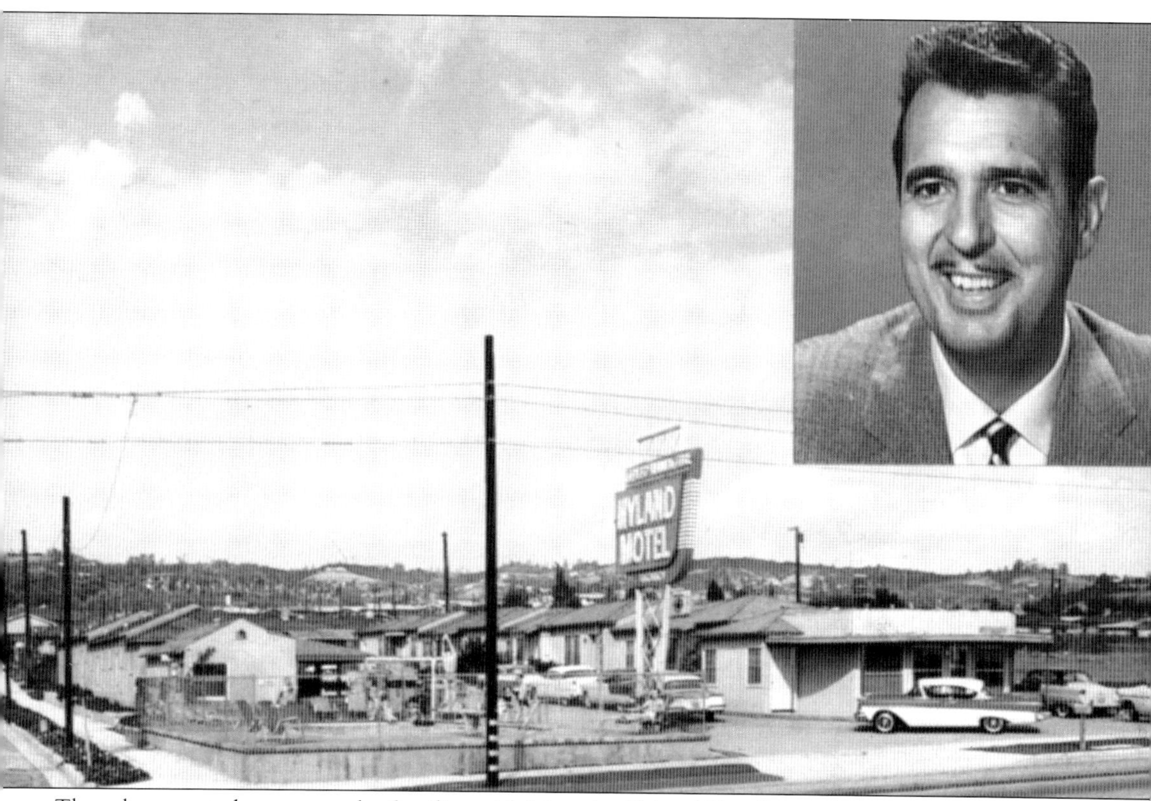

Though many today may not be familiar with him, the Hyland Hotel on Whittier Boulevard was once owned by a famous singer and television host of the 1950s–1970s, Tennessee Ernie Ford. Ford hosted his own variety show on NBC for five years, made guest appearances on *I Love Lucy*, and was a successful country and gospel singer. (Courtesy of the Thomas Pulley collection.)

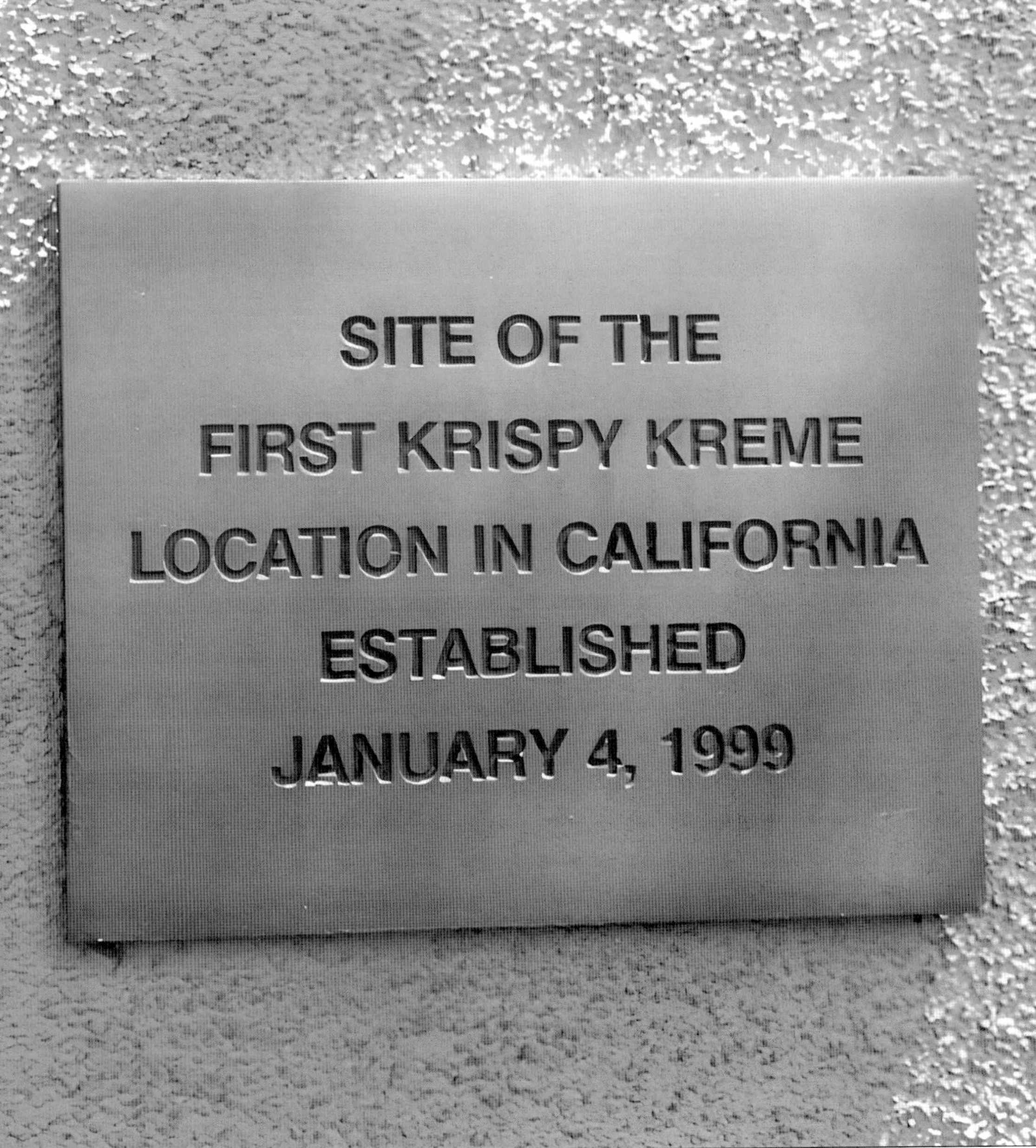

California's first-ever Krispy Kreme was opened in La Habra on January 4, 1999, at 1801 Imperial Highway, where Chick-fil-A is now located. It was reported that customers would form a line around the building by 5:30 a.m. for the doughnuts. Construction of a new Krispy Kreme is currently underway just under half a mile from its original location. (Courtesy of the City of La Habra.)

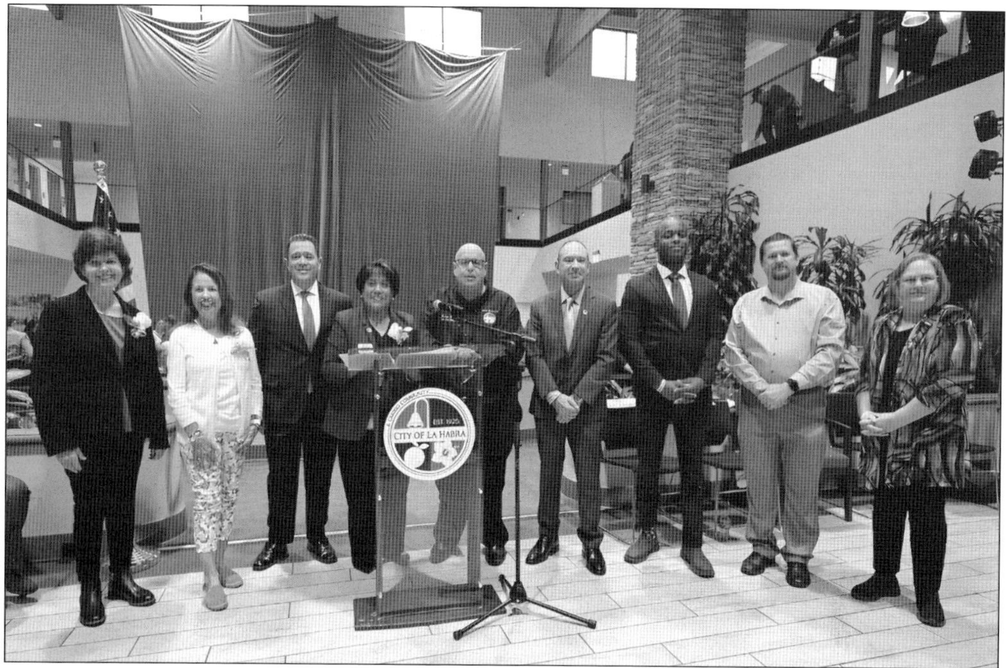

The La Habra City Council celebrated the city's 100th birthday with a community reception at city hall on January 21, 2025. Special activities included a Centennial Challenge Coin, a historical photograph exhibit, a presentation from family descendants of the early days, and the unveiling of the new mural. Special VIPs, including Cynthia Cramer Freeman, daughter of the late La Habra historian Ester Cramer, joined the current city council, including Mayor Pro-Tem Jose Medrano, Mayor Rose Espinoza, and Council Members James Gomez, Daren Nigsarian, and Delwin Lampkin, for the mural reveal. Below is the logo created in celebration of the anniversary year by the La Habra Centennial Committee. The logo includes nods to La Habra's history as a railroad and citrus town. (Both, courtesy of the City of La Habra.)

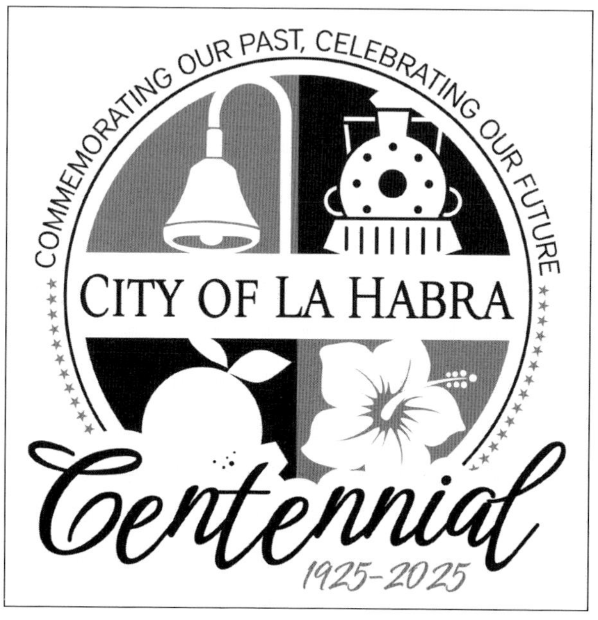

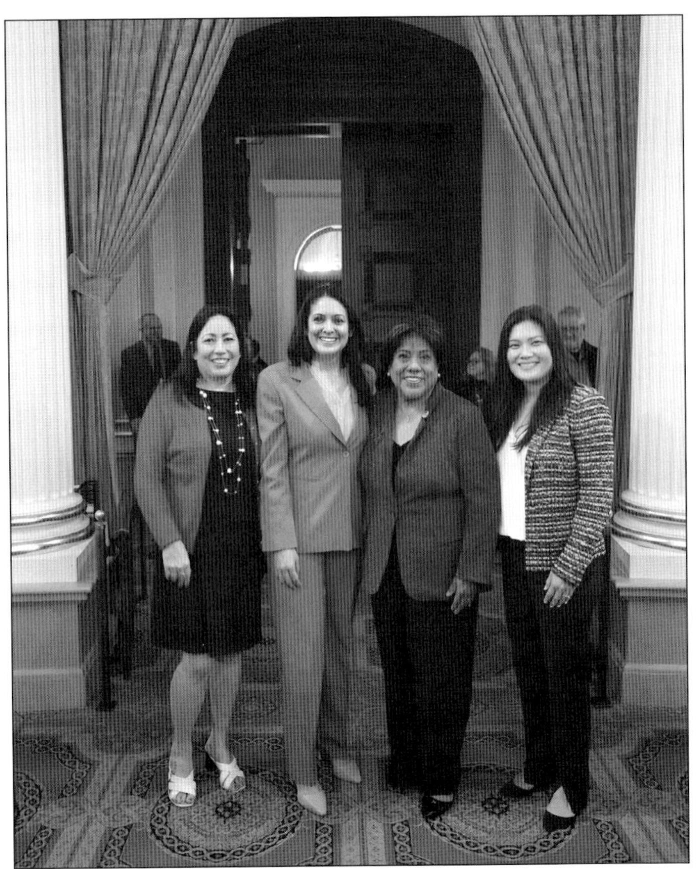

In recognition of La Habra's centennial anniversary, a delegation of city officials was invited to the California State Capitol in Sacramento on Thursday, January 23, 2025. Mayor Rose Espinoza, who is serving as mayor of La Habra for her fifth time since 2000, assistant city manager Gabriella Yap, and director of community services Kelly Fuijo represented the city for the occasion. From the Assembly floor, they accepted acknowledgements in honor of the city's historic milestone from Assemblymember Blanca Pacheco, who represents California's 64th Assembly District, pictured at left, and State Senator Tom Umberg, who represents California's 34th District, pictured below. (Both, courtesy of the City of La Habra.)

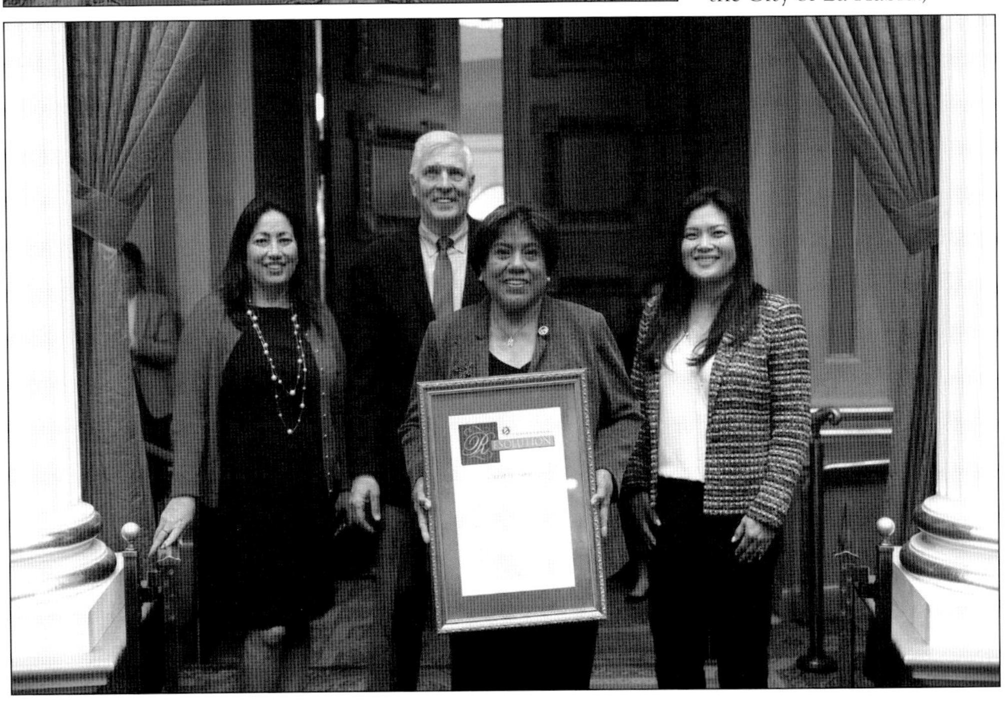

About the Committee and Authors

The City Centennial Planning Committee members are comprised of active and influential individuals in the La Habra community. They represent a wide range of organizations, including the Children's Museum at La Habra, Community Services Commission, Planning Commission, La Habra City Council, retired city employees, La Habra Host Lions Club, La Habra Collaborative, La Habra Art Association, La Habra Historical Museum, La Habra Woman's Club, La Habra City School District, Lowell Joint School District, Fullerton Joint Union High School District, La Habra businesses, and La Habra churches. Each member represented on the committee brings historical, institutional, and leadership value to providing a wealth of information for Images of America: *La Habra*.

Lauren Blazey is a second-generation La Habra native and grew up in the city. Her family, starting with her grandparents, has been living in La Habra since the 1960s, and her parents continue to be active members of the community. Her father, Michael Blazey, was on the La Habra City Council and Community Services Commission and is a member of the La Habra Historical Museum. Her mother, Jennifer Blazey, is on the Friends of the Children's Museum Board of Directors. Lauren graduated from Sonora High School before attending Whittier College for a degree in English literature, and recently, she earned a master's of fine arts from the University of Southern California's School of Cinematic Arts in Film and Television Production. She has collaborated with the city on several projects, including promotional videos for the recreation division, virtual concerts in the park during the COVID-19 pandemic, and photograph restoration for the La Habra Community Center lobby display.

Kimberly Albarian began her career in public service as an administrative intern in the city manager's office. She has worked for the City of La Habra for 19 years as the executive director at the Children's Museum at La Habra and as a community services manager and currently is the deputy director of community services. Kimberly has over 25 years of experience in recreation, grant writing and administration, fundraising and sponsorships, community collaborations, program development, and special events. Kimberly has a liberal arts degree in fine arts from Whittier College and a master's degree in public art administration from the University of Southern California. Kimberly resides in Tustin, where she was appointed to the Community Services Commission and was the chair of the first Public Art Commission.